HOW SCIENTIFIC PROGRESS OCCURS

INCREMENTALISM AND THE LIFE SCIENCES

ALSO FROM COLD SPRING HARBOR LABORATORY PRESS

A Cure Within

Davenport's Dream: 21st Century Reflections on Heredity and Eugenics

Evolution

Francis Crick: Hunter of Life's Secrets

Max Perutz and the Secret of Life

Mendel's Legacy: The Origin of Classical Genetics

Mutation: The History of an Idea from Darwin to Genomics

Times of Triumph, Times of Doubt: Science and the Battle for Public Trust

What a Time I Am Having: Selected Letters of Max Perutz

HOW SCIENTIFIC PROGRESS OCCURS

INCREMENTALISM AND THE LIFE SCIENCES

ELOF AXEL CARLSON

COLD SPRING HARBOR LABORATORY PRESS
Cold Spring Harbor, New York • www.cshlpress.org

How Scientific Progress Occurs
Incrementalism and the Life Sciences

Publisher and Acquisition Editor	John Inglis
Director of Editorial Services	Jan Argentine
Developmental Editor	Judy Cuddihy
Project Manager	Inez Sialiano
Permissions Coordinator	Carol Brown
Director of Publication Services	Linda Sussman
Production Editor	Kathleen Bubbeo
Production Manager	Denise Weiss
Cover Designer	Michael Albano

Front cover artwork: Artwork is from Flemming W. 1882. *Zellsubstanz Kern und Zelltheilung.* Verlag von F.C.W. Voge, Leipzig. Reproduced courtesy of Lilly Library, Indiana University, Bloomington, Indiana.

Library of Congress Cataloging-in-Publication Data

Names: Carlson, Elof Axel, author.
Title: How scientific progress occurs : incrementalism and the life sciences / Elof Axel Carlson.
Description: Cold Spring Harbor, New York : Cold Spring Harbor Laboratory Press, c2018. |
    Includes bibliographical references and index.
Identifiers: LCCN 2017059351 (print) | LCCN 2017061259 (ebook) | ISBN
9781621822981 (ePub3) | ISBN 9781621822998 (Kindle-Mobi) | ISBN
9781621822974 (case : alk. paper)
Subjects: LCSH: Biology--Research. | Science--Methodology.
Classification: LCC QH315 (ebook) | LCC QH315 .C2885 2018 (print) | DDC 570.72--dc23
LC record available at https://lccn.loc.gov/2017059351.

10   9   8   7   6   5   4   3   2   1

For a complete catalog of all Cold Spring Harbor Laboratory Press publications, visit our website at www.cshlpress.org.

Dedicated to the memory of
Claudia Sarah Carlson (1956–2016)
and John Gabriel Carlson (1962–2016)

Contents

Preface

In 1962, when Kuhn's *The Structure of Scientific Revolutions* first appeared, I was well into my career as a geneticist. My Ph.D. had been awarded in 1958, and four years later I had an active graduate program at UCLA. At that time, it was rare for geneticists to think about the philosophy of science, but it was not rare for scientists in any field to take an interest in the history of how their fields arose. As a student of H.J. Muller (1890–1967), I had taken his courses, which he approached historically—the formation of classical genetics as a series of battles—and these I recounted in 1966 in my first book on the history of science, *The Gene: A Critical History.* Although I was not persuaded that paradigm shifts as Kuhn portrayed them in 1962 applied to the history of genetics, I did not pursue my doubts as my career shifted to human genetics, teaching nonscience majors, and writing Muller's biography. It is thus 55 years later that I find myself writing this book. Neither the history of science nor one's own life is logical or predominantly predictable. I like to think that lives, careers, and the emergence of scientific fields, like evolution itself, owe much to opportunism.

I thank Christina Carlson for preparing the flowcharts. These charts help visualize some of the influences each new theory, experiment, or tool had in the formation of new fields in the life sciences. I benefitted from discussions on paradigm shifts with C.N. Frank Yang when he was at Stony Brook University and with Noretta Koertge at Indiana University. Judy Cuddihy, as usual, has been phenomenal in putting this book together and hunting for illustrations to grace the manuscript. Thanks to the editorial and production staff at Cold Spring Harbor Laboratory Press, Jan Argentine, Inez Sialiano, Carol Brown, Denise Weiss, and Kathleen Bubbeo, for their superb work.

ELOF AXEL CARLSON
Distinguished Teaching Professor Emeritus,
Department of Biochemistry and Cell Biology,
Stony Brook University, and
Visiting Scholar, Institute for Advanced Study,
Bloomington, Indiana
e-mail: ecarlson31@netzero.com

Introduction

For several decades, I have felt uncomfortable with the idea of paradigm shifts in the life sciences. In both my reading of Thomas Kuhn's *The Structure of Scientific Revolutions* (1962) and in the one conversation I had with Kuhn around 1974, in Princeton, when Bentley Glass asked me to look into setting up a history of science program at Stony Brook University, I asked Kuhn why paradigm shifts were rare or nonexistent in the life sciences. He told me that the physical sciences depended more on theory than biology and that biology was largely descriptive. He was also shifting his interest at that time to the vocabulary of science and how the words chosen to represent scientific findings and theories were influential in how science was perceived.

Kuhn's book was immensely stimulating and influential. I liked his idea of the shift that took place in Copernicus' model of the solar system, replacing the Ptolemaic view of an Earth-centered universe. I did not like his perception of what he called "normal science." It set up in my mind a group of scientists in which all but one, or a very few, perhaps over several generations, labored at putting pieces of jigsaw components into a mental picture that guided them. This was not really intended as trivializing the workers in normal science, but it certainly did not lift them to the esteem of the paradigm shifters who might be rarer than Nobel laureates by an order or two of magnitude. Creating a new worldview is extremely rare. Creating scientific revolutions is not. I believed that most of the Nobelists and high achievers I knew or whose work I admired introduced new discoveries, constructed brilliant experiments, introduced new tools to use, or developed new procedures that resulted in new fields of study. None of those were worldview constructions and none of those were paradigm shifts.

There are many ways to interpret how new fields arise or how they have evolved into their present states. The term "scientific revolution" suggests both an overthrow and replacement of how we interpreted the universe and applied it in our daily lives. No one argues against it having provided a pervasive influence on human life as the Renaissance shifted a transition from medieval to modern society. It ranks with the use of the "industrial revolution" to describe the era of steam-driven manufacturing. It also ranks with the term "agricultural revolution," an event spread out over millennia in prehistoric times. We would trivialize the term "revolution" if we also applied it to how safety matches were made or how seedless oranges could be perpetuated by grafting.

The term "theory" is also difficult to define and use because it can refer to an untested idea, a well-tested concept, or a settled fact. Is the atomic theory still a theory with the working out of 92 natural elements and a dozen or so synthetic ones? When do we elevate an idea from its status as a hypothesis to its status as a theory? Is it a theory or a fact that a forest is primarily composed of trees? Is it a theory or a fact that a living human body is composed of cells?

In this book I discuss theories, revolutions, and fields and how they are interpreted in the life sciences. If they are named, they can be described as paradigm shifts in Kuhn's 1962 sense. They can be described also as having incrementally developed through experimentation, new tools for generating new data, and insights or theories emerging from the abundance of new data. I do not believe it fair to either the idea of paradigm shifts or the idea of incrementalism to limit the terms "field," "revolution," or "theory" to just one of these two different ways to interpret the history of how they came into existence and their status today. I am aware that Kuhn reflected on the limits of his theory of paradigm shifts, and he believed his ideas were being extended into fields that did not apply (especially some of the social sciences and politics) and that his views were misinterpreted as denying the existence of an external reality or that all interpretations were constructions by consensus of those in any given field of knowledge.

As I studied the history of the gene concept, H.J. Muller's career when I did his biography, the history of classical genetics, the history of the idea of mutation, and the history of the biology of sex determination, I looked for paradigm shifts (in Kuhn's 1962 sense) and found none. Was I looking in the wrong place? Was I misinterpreting Kuhn? Was Kuhn correct that biology is still largely descriptive and that what we call the cell theory, the chromosome theory of heredity, the theory of the gene, the theory of evolution by natural selection, the theory of epigenetic development, and the theory of a molecular basis of life are just descriptive? If they are elevated to the status of paradigm shifts, what worldviews of science did they replace? What components in the older views were shuffled and renamed to create the new biological paradigm shifts?

This book is about the changes or progress in the life sciences that affect much of the basic science in these disciplines. I have presented narratives of the development of these fields, which are superficial to a scholar in the history of science who has paid a lot of attention to a richer story that could be told. Although I am aware of the many lesser-known contributors who fleshed out each field or each theory, I stress the major players. The life sciences are also connected to one another and it is impossible to isolate each discipline without reference to cognate fields and shared tools that they may use. That, too, is important to know about how the life sciences have evolved. I call the process incrementalism. It is Kuhn's "normal science" raised to a more significant level. Scientists are not solving a jigsaw puzzle. Most of the time they have no idea where innovation will lead, and the paradigm, if it exists, is a constantly changing one, not a photograph on a box propped up on the table for us to look at.

CHAPTER 1

Paradigm Shifts, Incrementalism, or Both?

What are the major attributes of life? The nature of the problem: Is it sudden Kuhnian paradigm shifts or is it an incremental change over time similar to the evolutionary process in which the past is modified with new knowledge and new techniques modifying a basic concept that is retained? Are both processes involved? What to look for to sort out the three possibilities.

Science as we know it begins in the 16th century with the work of Galileo Galilei and Renaissance scholars in Europe, especially in Italy, Great Britain, and Central Europe. Today we identify science with the union of technology, rational thinking, interpretation of data, a rejection of the supernatural as an explanation of observable phenomena, and experimentation to test or modify the implications of theories and inferences. Those components may have existed in isolated instances throughout human existence, but the Renaissance saw the first systematic effort to apply all of these to the physical and life sciences. If we contrast this modern science outlook that still prevails today with the knowledge of science in ancient times, we would see profound differences in how the world was seen. To those ancestors, the motion of the planets and the positions of the fixed stars were a source of information about oncoming misfortunes, good times, royal births, or sudden catastrophes. Astrology, not astronomy, was primary when studying the skies in an era that would not have telescopes for another two or three millennia. Mathematics was relatively secure as an exact science because of the rational nature of mathematical proofs. But even the Greeks, especially the Pythagoreans, saw in the patterns of numbers some connection, spurious as well as valid, to astrology and the arts (e.g., musical scales) and a possible divine aspect to their shapes or patterns. That belief in numerology later fed religious traditions (e.g., Kabbalah, Bible codes).[1]

In the life sciences there were scholars who studied the medicinal and commercial value of plant products. Herbals were part of the medical curriculum in the Middle Ages, with the bulk of medicinal products for treating the sick derived from plants. Also dyes for fabric came from plant products like woad or animal products like royal purple from snails along the Phoenician coast. The extraction, purification, and modification of techniques to make these effective was part of that ancient tradition of applied science, but most of the knowledge was passed through families involved in what later were called the guilds of the Middle Ages. Some of this knowledge was assembled by scholars like Aristotle and Galen. It is from those Greek and Roman sources that historians can grasp insights into how the ancients saw their world.[2]

Many saw their universe through the filter of their religious beliefs. This was certainly true of those who wrote the books of the Old Testament. Rainbows were signs or blessings or acknowledgements from God. Natural disasters, as we gather when reading Homer's *Odyssey*, were sent by deities who used humans to play out their divine rivalries and petty feuds. It was a god who churned the seas during storms. It was a god who tumbled rocks during an avalanche. It was a god who tossed lightning bolts or who crashed an immense anvil to create thunder. In the biblical tradition, it was transgressions against God that led to pestilences, plagues of locusts, and the parting of the seas to later smash together to destroy a hectoring army or to engulf the world in a flood. Some used sacrificed animal livers as auguries or means of predicting the future. Others, who were not guided by such direct religious sources for natural phenomena, created theories of health based on impurities or toxins that needed to be purged. Galen proposed four fluids or humors that he designated as blood, yellow bile, black bile, and phlegm. If they were in balance and not contaminated with toxins, the person was healthy. The ill had impurities or imbalances, correctible by bloodletting, purging, enemas, cupping, leeches, and other procedures. The functions of most organs in the body were unknown. There was no knowledge of cells, genetics, biochemistry, metabolism, or fertilization by gametes.[3]

How Is Knowledge of the Life Sciences Organized?

Today we organize a science in several ways. For the life sciences, the fields of botany (plant science) and zoology (animal science) made the first division. As noted earlier, botany was primarily a part of medicine. Zoology was separate from veterinary medicine or from farming with domesticated animals. There was no theory for why some animals can be domesticated and most remain wild and do not breed in captivity. Aristotle classified animals as warm-blooded and cold-blooded. He recognized four-legged organisms as a group. He studied embryos in chick eggs and concluded (correctly) that embryonic development was epigenetic—that is, form emerged gradually and was not simply a process of enlargement of a preexisting form. Form was somehow imposed on disorganized matter provided by the egg of the chicken. Aristotle believed the form to be present in the semen of the rooster, but he did not know how it worked.[4] The status of botany and zoology changed after the late 19th century when Louis

Pasteur, using a microscope and experimentation, showed microbial action in fermentation, putrefaction, and disease. Robert Koch established the techniques for studying these microbes and classifying them. The field of microbiology was added to universities.[5]

The attempt to classify shared and distinct features of organisms fell into the fields of anatomy and taxonomy. Anatomical study isolated the organ systems and determined some of their functions. The functional aspects that were not mechanical, like muscles and body movement, were shifted into a field called physiology. This was formalized by calling the structural basis of life morphology and the functional basis of life physiology. As microscopy was introduced to medical schools in the 19th century, the morphology of tissues became the science of histology. As stain technology and optics improved for microscopy, the details within the nuclei of cells led to another branch, cytology, which worked out cell division and gamete formation. Embryology also embraced microscopy in the 18th and 19th centuries and became a separate field.

In the late 19th and early 20th centuries, new fields emerged. Genetics studied heredity, especially through breeding analysis.[6] Ecology studied the relation of organisms to their environments and habitats. Systematics studied classification (taxonomy) in relation to evolution. Some fields combined to explore common interests. Thus, comparative anatomy wedded taxonomy to evolution.[7] Physiology spun off specialties like endocrinology, the study of hormones, in the early 20th century. The last fields to develop were biochemistry and molecular biology.[8] Biochemistry revealed biochemical pathways and thus fused genetics to biochemistry. Molecular biology fused the chemical and physical structure of molecules to their biological functions, especially after the discovery of DNA as genetic material that was physically organized as a double helix with an aperiodic sequence of nucleotides that made it the chemical basis of genes and their functions. This is just a sampling of dozens of specialty fields in the life sciences. There are probably about 50 such fields in health sciences and a similar number in the life sciences.

Do New Fields Emerge Suddenly or Gradually?

The purpose of my book is an exploration of those processes that lead to the emergence, fragmentation, union, and historical evolution of the life sciences. There are at least two major ways to interpret this process. In 1962, physicist and historian of science Thomas Kuhn (1922–1996) proposed a theory of paradigm shifts.[9] He classified most scientific work as "normal science." In this process, an initial theory or paradigm is an incomplete puzzle and the business of most scientists is filling in the unexplored parts of the theory and looking for a consistency when new additions are placed, much like a jigsaw puzzle analogy. When things do not fit or there are outright contradictions (called anomalies), the puzzle begins to collapse, a crisis ensues, and the old paradigm fails; a new theory rearranges the components into a new paradigm. The new paradigm solves the anomalies and a new meaning is provided to the old vocabulary. Kuhn called this process a "paradigm shift." His classic example was the shift from the Ptolemaic to the Copernican system in which the Sun shifted from its central status as a planet around the Earth into a central star, the Earth got displaced from the center of the then-known universe and became a planet, and the planets all revolved around the sun. Our moon shifted from being a planet around the Earth to a satellite around a planet. Note that in this Copernican paradigm shift the names and functions may change but the components are the same. Nicolaus Copernicus (1473–1543) did not add a new technology (the telescope was later introduced for astronomers by Galileo). Astronomy is not an experimental science. What changed was the mode of thinking about the relation of the components of the night sky.

In contrast to this way of seeing scientific revolutions, I propose designating normal science as "incrementalism."[10] In this model, change takes place in small (occasionally sudden or more significant) additions. So too are the pruning processes that eliminate outmoded observations and interpretations. Both the paradigm shift and the incrementalism models use the term "scientific revolution." For the paradigm shift model, the revolution is primarily a theoretical one. For the incrementalism model that I propose, the revolution is one of innovation through experimentation, new technologies, or the emergence, fusion, or splitting of fields of knowledge.

In the chapters of this book, I will discuss the two models in relation to different fields of the life sciences to see those that fit Kuhn's 1962 paradigm shift model, those that fit the incrementalism model, and those that are not readily

Thomas Kuhn received his B.S., M.S., and Ph.D. degrees from Harvard while studying physics. He shifted to the history and philosophy of science and taught at Berkeley, Princeton, and MIT. His most famous work appeared in 1962 as *The Structure of Scientific Revolutions*. Kuhn's book had a powerful influence on scholarly fields, especially in the social sciences where many added relativity of ideas and consensus as the basis for paradigm establishment. Kuhn tried to reject such claims. In popular idiom, the term "paradigm shift" has become a synonym for anything of intellectual, scientific, or social importance.

explained by either model or that combine features of both models. I will also discuss Kuhn's later views on how scientific fields arise.[11]

References and Notes

1. Neugebauer O. 1952. *The exact sciences in antiquity.* Princeton University Press, Princeton, NJ.

2. Nordenskiöld E. 1928. *The history of biology: A survey.* Scholarly Press, St. Clair Sholes, MI.

3. Porter R. 1997. *The greatest benefit to mankind: A medical history of humanity from antiquity to the present.* Harper Collins, New York.

4. Durrant M. 1993. *Aristotle's* De Anima *in focus.* Routledge, London.

5. Brock T. ed. 1999. *Milestones in microbiology.* ASM Press, Washington, DC.

6. Carlson EA. 2007. *Mendel's legacy: The origin of classical genetics.* Cold Spring Harbor Laboratory Press, Cold Spring Harbor, NY.

7. Osborn HF. 1894. *From the Greeks to Darwin: An outline of the development of the evolution of an idea.* Macmillan, NY; also see Gould SJ. 2002. *The structure of evolutionary theory.* Harvard University Press (Belknap Press), Cambridge, MA.

8. Olby R. 1974. *The path to the double helix.* University of Washington Press, Seattle.

9. Kuhn T. 1962. *The structure of scientific revolutions.* University of Chicago Press, Chicago.

10. Carlson, op. cit., p. 318, fn.

11. Dronamraju K. 1989. *The foundations of human genetics.* Charles Thomas Publishers, Springfield, IL.

CHAPTER 2

The Cell: From Empty Boxes to Coordinated Organelles

Hooke's boxes and buoyancy. Schleiden and Schwann with free formation of cells. The cell doctrine of Remak and Virchow. The nucleus, protoplasm, cell division, reproductive cells as gametes, mitochondria. Electron microscope, labeling for functions. Eukaryotic and prokaryotic organization.

In 1663, Robert Hooke (1635–1703) sharpened his penknife and carefully sliced a thin piece of cork. He placed it on what might be called the stage of his compound microscope and observed what looked like a miniature honeycomb. Because the cork was very light in color, he lifted off the slice and placed it on a piece of black plate. This time he observed the honeycomb was mostly empty space enclosed by what he called cells with walls. Hooke was a physicist primarily, and it immediately occurred to him that this could explain the spongy lightness of cork. "My microscope could presently inform me that here was the same reason evident that there is found for the lightness of froth, an empty honeycomb, wool, a sponge, a pumice-stone, or the like, namely, a very small quantity of a solid body, extended into exceeding large dimensions."[1] Hooke used the term "cell" because these little boxes reminded him of the monk cells frequently found in the courtyard of medieval churches and cathedrals. I recall the reverse experience when in 1973 I visited my father's childhood apartment house, Bellmansgaten, in Stockholm that faced a 14th century cathedral, Maria Magdalena Church. When I saw the boxlike monk cells against the iron gates, I thought of living cells and not dwellings for those who chose to commit themselves to a life of poverty and spirituality.

Hooke also correctly interpreted fossils, especially the foraminifera or empty calcareous shells of microscopic marine life, sea shells embedded in limestone, and petrified wood. He described these as once-living organisms that had been buried and their parts gradually replaced by mineral. He also used a telescope to study Jupiter and reported that it rotated on an axis because of the nightly movement of its Great Red Spot. He is best known to physicists for working out the linear relation associated with a spring and the weight it supports, or Hooke's law, which he described in words, "As the extension, so the

11

force," and as an equation $F = -kx$, in which F is the force, k is a constant that varies with the material used, and x is the distance by which the spring is extended.

Hooke published his findings in 1665 in a large book, *Micrographia*. He was a many-talented individual with an initial desire to be an artist; in fact, Hooke did the initial detailed drawings for the beautiful engravings in the book. His father thought that might be too uncertain a career. For most of his life, he was an employee of the newly founded Royal Society, which was committed to the advancement of knowledge through scientific observations and

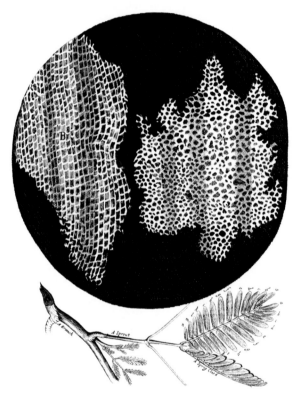

Robert Hooke described a slice of cork (*top*) taken from a small branch of a cork tree (*bottom*). He was struck by the boxlike structure of units that he called cells. Note, as he did, that they are empty and show the black background. The only function he assigned to them was buoyancy, because the woody matter was similar to a froth or a sponge in that the visible object mostly contains air. He published this illustration in 1665 in his book *Micrographia*. His microscopes were not powerful enough to see protozoa, bacteria, or similar small objects. It took considerable changes in lens making, stain technology, sectioning specimens, and a shift to microscopic anatomy in medical schools before a cell theory could emerge.

experiments. Hooke's job as Curator of Experiments was to repeat and show the experiments that were sent in as letters or short accounts to be published in the Royal Society Proceedings. A contemporary described Hooke as "the greatest mechanick this day in the world." Hooke was said to be very short, very thin, ugly, and somewhat misshapen, which made others uncomfortable in his presence. His personality was also abrupt and confrontational. He especially irritated Isaac Newton (1643–1727). Hooke claimed Newton stole his ideas on color theory and then published these and other findings without acknowledgement of his contributions. When Hooke died, Newton had almost every Royal Society item associated with Hooke removed to storage or discarded. There are no authenticated portraits of Hooke; the one known to have been painted was lost after the Newton housecleaning.

Leeuwenhoek Extends Microscopic Studies to "Animalcules"

A less complex microscope that consisted of a single spherical lens was prepared by Antonie van Leeuwenhoek (1632–1723) in Delft, the Dutch Republic.[2] He used melted beads from thin fibers of glass and inserted them individually in a metal plate. Leeuwenhoek made his living selling cloth and used his spare time for microscopy. He began sending letters and illustrations to the Royal Society in the 1670s. He wrote more than 500 letters, often including the specimens he used and illustrations in his own hand. Leuwenhoek discovered small animals that tumbled, swam, or propelled themselves by miniscule flagella or rows of beating cirri. Some were miniaturized forms similar to crustaceans. Some were

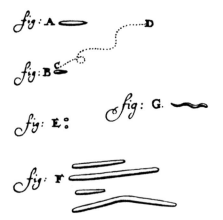

Leeuwenhoek described rod-shaped animalcules he scraped from his mouth in 1683. These may have been the species suggested by the protozoologist Clifford Dobell (1886–1949). They would have been too small to be seen in Hooke's microscope, but Leeuwenhoek's lenses could go up to 500 power, which is in the range of larger bacteria.

like small insects. Others were new to him, and he called all of them "animal-cules." At first, the Royal Society was skeptical that such microscopes were possible (his magnifications were ~250×–500×). The Royal Society sent investigators to Delft, and Leeuwenhoek showed them both his microscopes and his findings. He had each of the guests observe specimens with the microscopes. When they returned with their confirmation, Leeuwenhoek was made a member of the Royal Society.

What surprised Leeuwenhoek was the diversity and the wide distribution of these microscopic creatures. They were found in ponds, vases, saliva from his mouth, and even in his own semen. Some were less like animals than components of body fluids, like red blood cells. He did not use the term "cell," and he did not report a cellular structure of other forms of life. Although many of his illustrations suggest cellularity, he never drew a conclusion about the cellular composition of life. This was partly because many of his infusorians (especially protozoa) were single-celled organisms of complex structure.

Some believe he even saw bacteria, probably *Selenomonas* from his mouth. For those studying nature, another universe was revealed. There were myriads of tiny creatures that thrived in soil, streams, ponds, and virtually every fluid of the body. Telescopes revealed far more stars than the unassisted eye could see, and now microscopes were revealing an equally extensive universe of the very small creatures more numerous than the familiar life on Earth and previously not even imagined.

The Emergence of the Cell Theory

Equally gifted as an early microscopist was Marcello Malpighi (1628–1694), a physician who practiced near Bologna and who amassed considerable material on plant and animal microscopy. Like Leeuwenhoek, he sent his observations and drawings to the Royal Society in London, which published them in 1675 and 1679 as a book, *Anatome Plantarum*, or *Anatomy of Plants*. His work on capillaries in animals, *De Pulmonibus*, in 1661 provided the link between arteries and veins and the essential inferred connection for the circulation of blood that William Harvey (1578–1657) had shown by experimentation in 1628.[3]

The idea that familiar animals and plants were communities of cells began to emerge in the 1830s. Cells also shifted from being empty boxes to the contents within the boxes when cells were seen with better microscopes in the early 19th century. Robert Brown (1773–1858) noted a component within such filled cells that he called a nucleus. Brown was a versatile scientist who made his initial fame by going as a naturalist on a voyage to Western Australia in

1801 where he collected mostly new plant species (he named about 1200 plant species) and sent them back as preserved specimens. Some were potted and brought back by ship to the Royal Botanic Gardens at Kew. Brown did not extend his findings of nuclei in 1831 beyond the orchids and some monocotyledon plants he examined. He thought dicotyledons lacked nuclei in their cells.

Brown is better known for his discovery in 1827 of the random movements of small particles ejected from pollen in water, named in his honor as Brownian movement. He used pollen grains that broke open in water releasing starch granules and lipid particles that jittered about in an irregular or "swarming" motion. To make sure this was not something animated as a living component of the pollen, Brown used carbon dust and other mineral dust to repeat his observations and each showed the same Brownian movement. It was a physical phenomenon he had discovered.[4] In 1905, Albert Einstein applied the molecular theory of heat to liquids to explain Brownian motion and in so doing proved the existence of atoms. Atoms in liquid bombarded the suspended particles and caused their random zigzagging motion; his 1908 work suggested an experimental test that was confirmed by Jean Baptiste Perrin. It is an interesting example of biology leading to a main discovery in physics.

The inner composition of cells intrigued microscopists in the 1830s as they applied improved microscopes to living tissue. The idea of tissues dates to the French Revolution when Marie-François Xavier Bichat (1771–1802) used fine needles and a hand lens to tease apart components of organs. He did not use a microscope, which he considered a toy and too crude an instrument. He found some organs had membranes and some had fibrous composition like muscles. He called these organ components tissues. He identified 21 of these and argued that the body is a collection of a small number of tissues that compose the various organs. Bichat, unfortunately, died at the age of 30, probably of tuberculosis. The study of tissues would evolve into the field of histology when Bichat's tissues were examined by microscope.[5]

That shift came about in part from the studies of Felix Dujardin (1801–1860). He was originally an artist and watchmaker but took up microscopy and shifted his full attention to microscopic organisms. He was especially intrigued by foraminifera. Earlier, Leeuwenhoek, and others who followed, thought of these calcareous and siliceous shells as miniature mollusks. Dujardin used live foraminifera to show that these were single-celled organisms related to amoebas; they extruded their internal living mass through the pores and used these for locomotion or capturing prey. He called these filaments pseudopodia. There were no internal organs he could detect in this semisolid jellylike material that he called sarcode in 1835. He also noted that the sarcode in both the shelled and unshelled microbes that had pseudopodia also produced vacuoles with water or a clear fluid in them that would fill up and discharge.[6] In 1839, Jan Purkinje

(1787–1869) used the term "protoplasm" to describe what Dujardin had called sarcode. By the 1860s, protoplasm was popularized in Germany by Hugo von Mohl (1805–1872) and Max Johann Sigismund Schultze (1825–1874) and in Great Britain by Thomas Henry Huxley (1825–1895). Huxley in one essay described protoplasm as "the physical basis of life" (1868) and that virtually eliminated the further use of the term "sarcode" as the living material in cells.[7]

Improvements of Microscopes Accelerated Biological Findings

The rapid use of microscopes in the mid-19th century was a consequence of technological innovations. Spherical or curved lenses would bend light and, in some instances, break it like a prism into a rainbow of colors. Other lenses would focus the light into several planes. Both chromatic aberration and spherical aberration led to fuzzy images. This was not corrected until Joseph Jackson Lister (1786–1869), a wine merchant and amateur microscopist, worked out achromatic lenses using glass in which different salts were dissolved. He was the father of Joseph Lister (1827–1912), the founder of antiseptic surgery. Lead-containing lenses had a different focus than ordinary quartz lenses. By combining a layer of lead-bearing glass with normal quartz glass and polishing the lenses to the desired curvature, Lister found lenses that had a single focus and no chromatic distortion. He also developed a system of spaced lenses to correct these distortions.

Machines to cut very thin slices of tissue were developed in the mid-19th century. These microtomes produced a more uniform thickness of slices and could be adjusted for a range of thickness. Stain technology also developed when Joseph von Gerlach (1820–1896) discovered the first effective dye, aceto-carmine, to distinguish the nuclei and cytoplasm of cells as well as bring out the contrasts of the cells and their membranes. Gerlach had the good fortune of not being compulsive about a clean laboratory. He had left some slides that failed to take any dye and went home to his family. The next morning when he came in, he decided to look at his abandoned slides before washing them. He was stunned by the clarity of the dyeing process. He then worked out the ideal time for the tissue to sit in dye and the proper heating and other techniques to prepare permanent slides that could be used for reference. He found the dye worked slightly differently in different tissues, and very rapidly Bichat's tissues showed the visually distinguished cells that composed them. Histology became microscopic anatomy. The 21 tissues of Bichat became four major tissues—epithelial or lining cells that formed membranes, muscle cells with their peculiar banding patterns, nervous tissue that had long fibers or axons, and connective tissues or cells that were embedded in a matrix of cartilage, bone, tendon, or fat. Microscopy moved from the desks of amateurs at home to medical schools and the medical curriculum.[8]

Gerlach's findings in 1857 led to a rush of publications by other microscopists to work out techniques for effective staining and to test dozens of dyes. The most effective were new synthetic dyes, called aniline dyes, that were first synthesized by William Henry Perkin (1838–1907) in England at the youthful age of 18. Perkin's first dye was mauve, or aniline purple, a violet to purple dye that became the rage in the fashion industry in Victorian England. Perkin was trying to synthesize quinine and wound up with aniline purple. Perkin rapidly produced dozens of different dyes, but the older companies largely ignored his synthetic dyes and they stuck with their natural dyes, so he commercialized his discovery and opened a factory in 1857. By default, the dye industry shifted to Germany where Perkin's aniline dyes were produced in massive quantities and exported to the rest of the world.[9] Rudolf Peter Heidenhain (1834–1897) and others worked out multiple dyes on specific tissues, resulting in combinations like eosin and hematoxylin that dyed the cytoplasm pink (from the eosin) and the nucleus purple (from the hematoxylin). Both normal and diseased tissues were being studied and histology fed into the field of medical pathology for diagnosis of tumorous or diseased tissues.

The idea of organisms as communities of cells erupted in the 1830s and 1840s, but the most persuasive accounts were by Matthias Schleiden (1804–1881) and Theodor Schwann (1810–1882).[10] Schleiden published in 1838. Schwann published shortly afterward in 1839. Schleiden was one of the more remarkable biologists for his odd personality. He started out in law and became depressed because of his lack of success in that field, and in a depressed mood he shot himself in the brain. Fortunately, he survived with no serious damage to his mental or physical functions. He was impressed by the care he received and decided to shift to medicine. That too did not pan out so he switched to botany in which he had remarkable success. He popularized botany (*The Plant—A Biography*), and as he applied his microscope, he found that every major group of plants he studied had a similar composition of cells. He recognized the cell as the structural unit of plant life. In one account of his career, he claimed he was traveling by train and happened to meet a fellow physician, Theodor Schwann, a Belgian. When Schleiden mentioned his ideas of plants as being communities of cells, Schwann pulled out his working manuscript that showed animals are communities of cells. The somewhat confrontational and erratic personality of Schleiden happened to work well with the relaxed, good-humored, and accepting personality of Schwann, who struck up a friendship that prevailed the rest of their lives. They agreed to publish their work and promote their cell theory, as they called it. All organisms, they claimed, are composed of cells.

But how did these cells form? Schwann had a chemical background (he coined the term "metabolism") and suggested to Schleiden that cells might be likened to crystals. Crystals precipitate from supersaturated solutions and perhaps the nuclei, then called cytoblasts, were such early cells in the making. This made good sense to Schleiden and they added to their cell theory the proposal that there is a "free formation of cells" analogous to crystallization. The cell walls or membranes would form around the enlarging nuclei and establish new cells. Essentially, new plant cells formed from the nuclei of old plant cells and the original wall of the cell dissolved allowing the newly formed cells to enlarge.

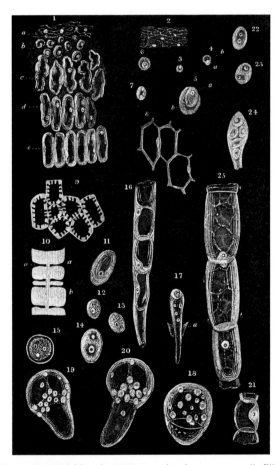

Cells with nuclei, seen by Schleiden, in 1838. Here he shows some cells filled with fluid (*8*), some with multiple nuclei (*18*), and some in the act of partitioning (*20*). Some show what he and Schwann called free formation in the process of forming metabolic crystals (*15*).

The Cell Theory Becomes the Cell Doctrine

By the 1850s, the cell theory was well established, but the free formation of cells was not. That doubt was expressed by Robert Remak (1815–1865) who believed cells produced cells by a process of cell division.[11] He claimed that a single cell partitions into two cells and the cell membrane plays a role by pinching in to separate the two cells formed by cell division. Remak was born in Poznań, today part of Poland and then part of Prussia. He was a Jew who did not convert. This allowed him to get an education as a physician but not a position as a professor until 1859. He had to make a living through private practice and to do his research and publish without academic support or recognition.

Unfortunately for Remak, his ideas were incorporated by Rudolph Carl Virchow (1821–1902) without acknowledgement. Virchow came to the same conclusion and promoted this idea as a "cell doctrine"—*omnis cellula e cellula*, or all cells come from cells. Virchow came to his modification of the cell theory from his interest in pathology. As a young physician, he went to Silesia to study the high death rate of miners in Silesia. He wrote up a report in 1848 that the major cause was low wages, which did not go well with the government agency that sent him. In addition, his timing was bad; the Socialist revolutions were breaking out throughout Europe. Virchow was suspended and went to a western German state to study pathology. He began using the microscope and noted that cancer patients had uniquely different cells from the tissues in which they were embedded. He inferred that each tumor enlarged and had its origin from a single cell. It was this working backward method that enabled Virchow in 1855 to establish the origin of cells from preexisting cells, including tumors, which would have some event that shifted a normal cell into a tumor cell.

An important implication for Virchow was that all of life had to go back to the first cell from which life arose. For Virchow that was an act of a Creator. He felt the cell doctrine repudiated the long-held belief in the spontaneous generation of life.[12] For Darwin some 10 years later, something analogous to a free formation of cells was used, establishing, without assistance from a supernatural source, a first cell from which all life would descend. Virchow founded the first journal for cellular pathology, now known as *Virchows Archiv*, and promoted his cell doctrine to his medical students and in his many publications. Remak was embittered that his work was ignored or not credited, although Virchow in 1868 acknowledged that priority for Remak as being the first to propose the cell doctrine. It is rare to change the way we learn our history of science, but simple justice should refer to the Remak–Virchow cell doctrine, just as the Hardy law became the Hardy–Weinberg law some 50 years after it was independently published by Hardy in England and Weinberg in Germany.[13]

The Rise of Cytology and the Process of Cell Division

The description and interpretation of cell division shifted histology to cytology. In the 1850s, Remak and Virchow saw cells and nuclei and the surrounding cell membrane that held the protoplasmic contents of the cell. The story of cell division shifted when the advances of stain technology revealed the presence of chromosomes forming in the nuclei of cells. Albert von Kölliker (1817–1905), a Swiss zoologist, in 1862 distinguished the protoplasm around the nucleus as cytoplasm. He applied the new stain technology and developed ways to harden the tissues and section them, especially to study embryo formation in mollusks and later in vertebrates. He noted earlier, in 1850, the presence of structures in muscle cells that he called "sarcosomes." He returned to these later and found they could separate from the muscle fibers and that they had their own membrane. It would be another 40 years before these were renamed as mitochondria. During the 1860s, there was a growing belief that the nucleus must play some significant role in the cell's functions. With the exception of single-celled organisms found for some algae, fungi, and protozoa, the known phyla of animals and plants were multicellular and composed of cells. Virtually all the cells had nuclei. Also in the 1850s and 1860s, it looked likely that fertilization involved the union of two reproductive cells. In vertebrates, these were a sperm and an egg. Proof of this belief would require more time and the details would be spread over two or three decades, but Ernst Haeckel (1834–1919) noted that cells varied in size from massive yolk-filled eggs to tiny animalcule-like sperm that swam with a vibrating tail. Although the cells had an immense range of sizes, their nuclei did not. In most of the tissues of an organism, the nuclei were of identical size. Haeckel in 1865 inferred from this that the nuclei, and not the cytoplasm, were more likely to contain the hereditary basis for the species' characteristics. He assumed that the hereditary contributions of the male and the female would be equal.[14]

Weismann Proposes a Theory of the Germplasm

August Weismann (1834–1914) predicted the existence of a special cell division for the formation of reproductive cells that would reduce the hereditary content by half. Weismann's major contribution to biology was his theory of the continuity of the germplasm in 1883 (i.e., only changes in the germplasm could be inherited).[15] He believed that this was set aside early in embryonic development and that somatic tissue (body tissue) was transient and died off during the life cycle. He based this on his doctoral work in 1875 in which he examined insect embryos and noted an early isolation of the future germ cells isolated in a "polar cap" at one end of the embryo. It was this preserved germinal material that later in the adult matured and produced offspring through fertilization by gametes. This reversed the time direction for cell theory. Cells had a past from a preexisting cell, but Weismann's theory of a separate germplasm shifted the direction of time to the future. It was the germplasm that had continuity as long as the species existed. The somatoplasm was limited to a single generation only to die out as age, misfortune, or disease claimed that individual. But the future was guaranteed through the germplasm. Also, part of Weismann's theory of the germplasm was his rejection that changes in the somatic tissues could have an influence on heredity (i.e., inheritance of acquired characteristics), refuting Lamarck. He showed by doing experiments with mice (a total of 901 mice produced by five generations) that tail length was innate and not influenced by surgical amputation or mutilation.

The Chromosome Number Becomes Important for Cytology

Mitosis was worked out in the late 1870s. Heinrich von Waldeyer-Hartz (1836–1921) described the string-like objects that replaced the nucleus and named them chromosomes in 1882. But it was several years earlier that Walther Flemming (1843–1905) worked out a sequence in which the nucleus began forming clumps and thick strands instead of a fine network of fibers called chromatin. Those strands filled the space of the nucleus as the nuclear membrane disappeared. The first phase was later called prophase. It was followed by an alignment of the chromosomes in the equatorial region of the cell. This state later acquired the name metaphase. This was followed by the movement of the chromosomes to opposite poles of the cell. Each half of the cell got the same chromosome number as it had at prophase, suggesting the chromosomes are longitudinally divided during mitosis. This stage of separation and early movement was called anaphase. The last stage followed the chromosomes to the poles and a pinching movement in the cytoplasm would partition the cell into two cells, each cell having a new nucleus forming as the chromosomes unwound to their chromatin state. The last stage was eventually called telophase.

Much of the cytological analysis of reduction division for reproductive cells was performed by Belgian biologist Edouard van Beneden (1846–1910). He was much influenced by Flemming's finding of mitosis, and he attempted to refine the changes taking place during the stages of mitosis. He also chose a fortunate species for his studies, the nematode worm *Ascaris megalocephala*.[16] He obtained one species with a chromosome number of 4 in the somatic tissues and a chromosome number of 2 in the mature gametes. He noted that during fertilization, the nucleus derived from the sperm and the nucleus derived from the egg fused to produce a new nucleus with a restored chromosome number of 4. This implied something more significant. If the chromosome number is preserved from parent to offspring by going through Weismann's predicted reduction or halving, then this should be true for other species. This led to a survey of several species by van Beneden; in 1887, he drew two findings from his work. First, there was a constancy of chromosome number from parent to offspring and in all the somatic tissues of the organism. Second, there was a constancy of chromosome number for a given species, with all its members showing the same chromosome number. Mitosis guaranteed the chromosome number from cell division to cell division in the somatic tissues. Meiosis, the special reproductive division that van

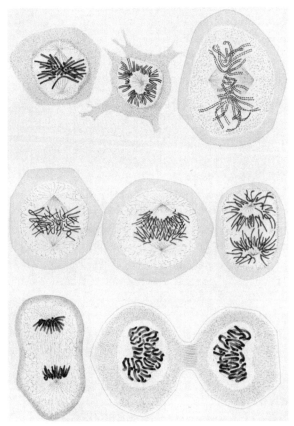

Walther Flemming first recognized that the chromosomes (described and named by Waldeyer) formed a set of unusual distributions. Eventually these acquired the names prophase, metaphase, anaphase, and telophase. After the cell sent the chromosomes to opposite poles of the cell (*lower left*) in telophase, the cell partitions into two by a process called cytokinesis (*lower right*). Metaphase is represented in the *middle left* and anaphase movement begins with the central cell (*middle right*). Not shown is interphase, when the cell is not actively dividing. Prophase is seen at the *upper left* where the chromosomes have condensed into visible threads. Flemming's illustration is from 1882.

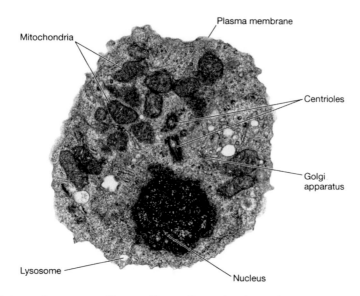

The electron microscope readily magnifies a cell an order of magnitude greater than optical microscopes. Unlike stain technology for making slides, the specimens are impregnated with metallic salts and then beams of electrons passing through are spread out and exposed to a fluorescent screen from which a photograph can be taken. In this eukaryotic cell at magnification of 15,000×, the nucleus and cytoplasmic contents are labeled.

Beneden analyzed and named, guaranteed the same chromosome number for the next generation through a process of fertilization.

Cytoplasmic Organelles Revealed by Microscopy and Physiological Chemistry

The relation of chromosomes to heredity will be discussed in later chapters, but the components of the cell shifted again from the nucleus to the cytoplasm. In 1888, Theodor Boveri (1862–1915) worked out the relation of the centriole, the formation of the mitotic spindle, and the centrosomes during cell division. A year later, Camillo Golgi (1843–1926) was examining the brain of a barn owl and he identified in some neurons a bunching of membranes near the nucleus that became named for him as the Golgi apparatus.

In 1889, the first biochemical aspect of the cell was identified by Richard Altmann (1852–1900), who described the differences in staining between nuclei and cytoplasm and associated the highly acidic material that was present in the nuclei as nucleic acid. Altmann coined the term "nucleic acid" to replace the term "nuclein," which had earlier been described and named in 1869 by a Swiss physician Johannes Friedrich Miescher (1844–1895), who was in the laboratory of Ernst Felix Hoppe-Seyler (1825–1895) at the University of Tübingen. Miescher thought that proteins were the "hereditary molecules." He was looking for phosphorus-rich compounds and used pus cells because they were abundant and easy to extract from hospital bandages. He chemically characterized nuclein as a phosphorous-rich substance and isolated the nuclei from the cell cytoplasm and the nuclein from the suspension of associated proteins by using an alkaline extraction followed by an acidic precipitation. This was

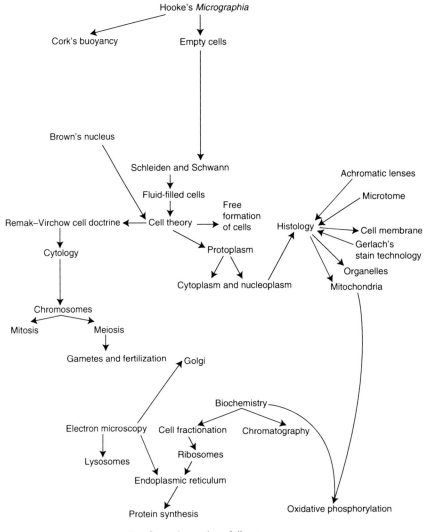

See figure legend on following page.

a new technique and a pioneering effort to apply chemistry to biology through isolated cell components studied chemically in purified form.[17] It would lead to the formation of a new field of cell biology. In 1924, Miescher's nuclein was identified by a specific stain (the Feulgen stain) and confirmed to be DNA or deoxyribonucleic acid that was present in the nuclei, particularly the chromosomes. Miescher and Albrecht Kossel (1853–1927) both were in Hoppe-Seyler's laboratory about the same time. Kossel took an interest in nucleic acid and began isolating its component chemicals. From 1885 to 1901, Kossel isolated and named five nitrogenous compounds that he called adenine, guanine, cytosine, thymine, and uracil.[18]

Electron Microscopy Amplifies Cell Organelles and Cell Structure

The morphology of cells shifted dramatically in 1931 when Ernst Ruska (1906–1988) introduced the electron microscope.[19] The idea came to him as a graduate student, and he developed an electron lens using a magnetic coil to spread out electrons as they passed through or around small objects and increased their size to 10,000 or more times their original size. He spent 1931 to 1939 developing a commercial model of his electron microscope. He was at Fernseh Ltd. from 1933 to 1937 and at Siemens-Reiniger-Werke AG from 1937 to 1956, where he developed the Siemens Super Microscope in 1939. His brother

---

Hooke's *Micrographia* (1665) introduced the microscope as a tool for studying living things and their components. His observation of a slice of cork revealed what he called "cells," and he associated their function as providing buoyancy. He did not offer a theory of cells as building blocks of living matter—his cells were empty. Schleiden and Schwann (1835) offered a cell theory that animals and plants were composed of cells. They assumed there was a free formation of cells, like crystals forming from supersaturated liquids. With the introduction of achromatic lenses, cells were shown to be filled with a fluid material and to contain a nucleus. Gerlach's introduction of stain technology (1857) led to the identity of the cell's fluid as protoplasm and a field of histology in which tissues were analyzed into their cell types. Stain technology also revealed cell organelles, with mitochondria and the cell membrane being the first to be assigned such status. Remak and Virchow argued that all cells arise from preexisting cells and they called this the cell doctrine. It led to the field of cytology and the discovery of chromosomes and their role in mitosis and meiosis. Meiosis led to the recognition of germ cell formation and the formation of a zygote through fertilization of one egg and one sperm. Biochemistry grew out of organic chemistry and organelles were studied for their chemical composition and functions. The tools of cell fractionation, chromatography, and radioactive labeling were particularly effective in working out these components and functions. This led to the association of oxidative phosphorylation with mitochondria, protein synthesis in the endoplasmic reticulum, and intracellular digestion in the lysosomes. The electron microscope also played a major role in identifying the structure and location of these organelles in the cell.

Helmut Ruska (1908–1973), a physician, used the electron microscope to study cells and microorganisms. This jumped the observable size of the cell by several orders of magnitude, and a fine detail was revealed. The cell membrane was shown to be a bilayer that was chemically identified as mostly lipid. The mitochondria turned out to have both an outer membrane and a much-folded inner membrane. In 1929, the molecule adenosine triphosphate (ATP) had been identified by Karl Lohmann (1898–1978), and the manufacture of that molecule was associated with the mitochondria and the production of chemically portable energy to drive the metabolism of the cell. The biochemical function of ATP and its role in mitochondrial function was worked out by Fritz Albert Lipmann (1899–1986) between 1939 and 1941. It was Lipmann who coined the phrase "energy-rich phosphate bonds" as the essential feature in ATP that permitted energy to circulate in the cell in a chemically stored form.

The bulk of the cytoplasm was filled with folded layers called the endoplasmic reticulum, which were later shown to be involved in protein synthesis, a complex story that will be discussed in Chapter 6. Additional organelles of the cell were identified in the 20th century. The chloroplasts and their photosynthetic layers were worked out in plant cells. Lysosomes were shown to be digestive organelles with as many as 80 different enzymes that recycled cell wastes. The nucleus turned out to have not a membrane like the outer cytoplasm but an envelope perforated with pores. In the late 20th century, the cytoplasm was shown to have a cytoskeleton. This story is ongoing; although cells can be reconstituted from the isolated parts taken from other cells, the organization of even the simplest cell cytoplasm, called the cytosol, is not yet possible. Many cell processes are involved and some kind of cytoskeletal organization occurs, but a full inventory and understanding of its role in the cell cycle is likely to emerge in the near future.[20] The 20th century, as these findings suggest, can be described as the time when morphology and physiology were united at the cellular level. From Harvey's analysis of the pumping action of the heart to generate blood circulation, it took about 250 years to show a similar relation of form to function at the cellular level for cell division and 300 years from Harvey's breakthrough to see such functions related to cell organelles and their biochemical activities.

Tracing the Evolution of a Field

This brief account shows how the idea of the cell shifted from an empty box to a container of organelles that performed the major functions of life by numerous contributions from 1665 to the present. Some scientists supplied components, some reshaped the original contribution of another colleague, and some replaced erroneous interpretations or faulty observations. The path is not quite a drunkard's path, but it does zigzag and it backpedals at times, with a direction toward complexity as new technologies provide new opportunities to explore, to confirm, and to add to the growing sense of what cells are and what they do. The major findings, or revolutions, are new tools and techniques (e.g., microscopes, stain technology, biochemical analysis of cell organelles) as well as theories (e.g., the cell theory, the cell doctrine, continuity of the germplasm, reduction division) and the important contributions of observation (the details of mitosis and meiosis). What is missing or hard to identify are paradigm shifts as Kuhn perceived them in 1962. There is no shuffling and renaming of parts coordinated into a new theory. Instead, there is incrementalism, hundreds of additions, tweaks and refinements, competing interpretations, and old ideas discarded as new data floods into a growing field that keeps spinning off new specialty fields and making connections to other fields.

References and Notes

1. Hooke R. 1665. *Micrographia: Or some physiological descriptions of minute bodies made by magnifying glasses. With inquiries and observations thereupon.* J. Martyn and J. Allestry, London. See passage, "Of the schematisme or texture of cork, and of the cells and pores of some other such frothy bodies," in 1967. *Modern biology* (ed. Carlson EA), pp. 19–21. Braziller, New York.

2. Dobell C, ed. 1932. *Antony van Leeuwenhoek and his "little animals": Being some account of protozoology and bacteriology and his multifarious discoveries in these disciplines.* Constable and Company, Limited, London.

3. Malpighi M. 1675, 1679. *Anatome plantarum.* Royal Society, London. For an account of Malpighi's life and contributions see Adelmann H. 1966. *Marcello Malpighi and the evolution of embryology,* 5 volumes. Cornell University Press, Ithaca, NY.

4. Brown R. 1866–1868. *The miscellaneous botanical works of Robert Brown.* Hardwicke, London.

5. Bichat X. 1816, 1821. *Traité des Membranes en Général et de Diverses Membranes en Particulier.* 1816. Méquignon-Marvis, Paris. Trans. Houlton J, *A treatise on the anatomy and physiology of the mucous membranes; with illustrative pathological observations.* 1821. Printed for J. Callow, London; 1827.

6. Dujardin F. 1841. *Histoire Naturelle des Zoophytes. Infusoires, Comprenant la Physiologie et la Classification de ces Animaux, et la Manière de les Étudier à l'Aide du Microscope.* Roret, Paris.

7. Huxley T. 1868. On a piece of chalk. *MacMillan's Magazine* 18: 396–408 (reprinted 1967, Scribner's, New York). Also available online at https://mathcs.clarku.edu/huxley/CE8/Chalk.html. Blinderman C, Joyce D. 1998. *The Huxley file.* Clark University, Worcester, MA.

8. Conn HJ. 1933. *A history of staining.* Book Service of the Biological Stain Commission, Geneva, NY.

9. Garfield S. 2000. *Mauve: How one man invented a colour that changed the world.* Faber and Faber, London.

10. Schleiden MJ. 1839. Contributions to phytogenesis, pp. 229–265. In Schwann T. *Mikroscokpische Untersuchungen über die Uebereinstimmung in der Struktur und dem Wachsthum der Thiere und Pflanzen*. Sanders, Berlin. Trans. Smith H. 1847. *Microscopic researches into the accordance in the structure and growth of animals and plants*. Sydenham Society, London.

11. Remak R. 1852. "Ueber extracellulare Entstehung thierischer Zellen und über Vermehrung derselben durch Theilung." *Arch Anat Physiol Wissenschaftl Med* **1**: 49–50.

12. Virchow R. 1858. *Die Cellularpathologie*. Translated into English 1859 as *Cellular pathology* and reprinted 1978. Churchill, London. For a biography see Ackerknecht EH. 1953. *Rudolf Virchow: Doctor, statesman, anthropologist*. University of Wisconsin Press, Madison, WI.

13. Stern C. 1943. The Hardy–Weinberg law. *Science* **97**: 137–138.

14. Richards R. 2008. *The tragic sense of life: Ernst Haeckel and the struggle over evolutionary thought*. University of Chicago Press, Chicago.

15. Weismann A. 1892. *Das Keimplasma. Eine Theorie der Vererbung*, Fischer, Jena, Germany. Trans. Parker WN, Rönnfeldt M. 1893. *The germ-plasm: A theory of heredity*. Scribner's, New York. For a biography of Weismann, see Churchill F. 2015. *August Weismann: Development, heredity, and evolution*. Harvard University Press, Cambridge, MA.

16. For a collection of English translation papers on the cytological studies leading to the discoveries of mitosis, meiosis, and the chromosome theory, see Voeller BR. 1968. *The chromosome theory of inheritance*. Appleton-Century-Crofts, New York. Also see Flemming W. 1878. Zur Kenntniss der Zelle und ihrer Theilungs-Erscheinungen. *Schr Naturwiss Ver Schlesw-Holst* **3**: 23–27.

17. Dahm R. 2008. Discovering DNA: Friedrich Miescher and the early years of nucleic acid research. *Hum Genet* **122**: 565–581.

18. Kossel A. 1881. *Untersuchungen über die Nukleine und ihre Spaltungsprodukte [Investigations into the nucleins and their cleavage products]*. Trübner, Strassburg, Austria.

19. Freundlich MM. 1963. Origin of the electron microscope. *Science* **142**: 185–188.

20. Gall J, McIntosh JR, eds. 2001. *Landmark papers in cell biology: Selected research articles celebrating forty years of the American Society for Cell Biology*. Cold Spring Harbor Laboratory Press, Cold Spring Harbor, NY.

CHAPTER 3

The Theory of the Gene: From Abstract Point to Nucleotide Sequence

Hereditary units in the 1860s, Mendel's factors, Darwin's gemmules, de Vries' intra-cellular pangenesis, Bateson's unit characters, Morgan's factors, Johanssen's genes. Genes as points on a map, genes as lengths of chromosomes, position effects, deletions or line mutations, multiple alleles, pseudoalleles. Fine structure of genes, the gene as DNA, the gene as a nucleotide sequence, split genes. Regulating genes for transcription and translation, silencing and regulating genes with epigenetics.

The term "gene" was introduced by Danish geneticist Wilhelm Ludvig Johannsen (1857–1927) in 1909, but the idea of hereditary units goes back to the 1860s when Gregor Mendel (1822–1884), Herbert Spencer (1820–1903), and Charles Darwin (1809–1882) introduced that concept.[1] For Mendel, trained to be a science teacher in his monastery in Bruno in the Czech state of the Austro-Hungarian Empire, these were "factors" that provided discontinuous inheritance with some factors being recessive and some dominant. His experiments revealed how they were transmitted and what would be expressed in a multigeneration series of crosses.[2] For Spencer, they would be theoretical units larger than known molecules and smaller than cells or the nuclei of cells. He called them "physiological units" and assumed that in some unknown way they determined the development, structure, and composition of living things. Spencer was not a scientist. He was a philosopher and his concept was a Baconian inference based on a universal atomism in which everything was composed of smaller hierarchies of units. Spencer was also an autodidact and an early proponent of what would later be called the Libertarian movement. In his lifetime, he was seen as a crank, a bore, a Socialist, a feminist, a laissez-faire capitalist, an ardent advocate of degeneracy theory, a forerunner of the eugenics movement, and a subversive who saw the state as the enemy of the individual.[3]

Charles Darwin, of course, was the evolutionist whose theory of natural selection upended world science and forced a conflict between science and religion that had simmered on a mental back burner since Galileo was put to house arrest for his endorsement and demonstration of the Copernican model of the solar system. Darwin's view of hereditary units was promoted in 1863 as a "provisional theory of pangenesis." Darwin believed that cells extruded small hereditary units—Darwin called them "gemmules"—and these were gathered in the reproductive cells and transmitted one's heredity as it was at that moment. It was a shifting heredity responding to the environmental shaping of the body's morphology and physiology.[4]

The Early Tests of Hereditary Units

Of these three units proposed for heredity, Mendel's findings remained obscure until they were revived in 1900 and Spencer's were dismissed as vague speculation of no practical benefit to breeders or scientists. Darwin's were shot down by his cousin Francis Galton, who put the model to a test using transfusions of blood in rabbits of different pelt color. His results showed no evidence of transmission to the reproductive cells of those rabbits receiving blood with the alleged gemmules of a different variant in them.

Darwin's gemmules were similar to older ideas by Greek philosophers, who speculated that such units somehow formed and got concentrated in the semen or in the female component for the next generation. There was no cell theory in antiquity, and there were no imagined mammalian eggs or spermatozoa. Older ideas were based on form imprinting itself on formless matter, with the formative element being male semen and the responsive shapeless mass (often menstrual blood) being the female's contribution, sometimes called female semen.

Hugo de Vries (1848–1935) revived Darwin's model in 1892.[5] The generation after Darwin introduced natural selection as the major biological theory of the 19th century, and cells acquired a new status as the basic units of all organisms. The nuclei of cells were seen as the source of hereditary units. August Weismann's studies refuted a Lamarckian interpretation of environmental modification of these units. Instead, de Vries argued, the units remained within the cells. Variation was a consequence of a shuffling of these units and occasional modifications of them by unknown means. He called his model "intracellular pangenesis," and he discarded the term "gemmules" and replaced it with his tribute to Darwin; they were "pangenes."

de Vries said he turned to plant breeding because he was looking for pangenes. This led to his being one of three continental scientists who rediscovered or confirmed Mendel's findings depending on when they first read Mendel's paper in a literature search. de Vries, Carl Correns (1864–1933), and Erich von Tschermak (1871–1962) get that honor for their 1900 rediscovery papers that established Mendel as the pioneer founder of the study of heredity by breeding analysis. In the immediate flurry of confirming papers and new extensions of

Mendelism to other plants and to animals, none of the investigators acknowledged de Vries's pangenes as the units involved. They either used the Mendelian term "factor" or, like William Bateson (1861–1926), introduced them as "unit characters."[6]

By 1909, the lack of a standard name and the misconceptions attributed to the various terms led Johannsen to lop off the prefix from de Vries's term; he called them "genes."[7] This honored both Darwin (pangenesis) and de Vries (pangenes), but Johannsen added an important clarification. He said no one knew what genes were or what they did and the term should be undefined in chemical, physical, or physiological terms, except to connote that it was a unit of inheritance. It took about 5 years before the competitive terms disappeared from most scientific articles. Johannsen had earlier in 1903 coined the terms "genotype" and "phenotype."

Morgan's Fly Laboratory Shifts Hereditary Units to a Theory of the Gene

In the United States, Thomas Hunt Morgan (1866–1945) and his students in the Columbia University "Fly Room" promoted the theory of the gene. They

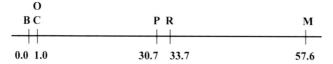

In this first linkage map, ordering six X-linked genes found by Morgan's fly laboratory, Sturtevant used the letters B for normal body color (b being yellow body color), C for normal eye color (c being white eye), O for the normal gene for the eosin variant, P for the normal gene for the vermillion eye color (p), R for the normal counterpart to miniature wing variant (r), and M for normal gene for rudimentary wing variant (m). By 1920, Muller introduced a standard nomenclature for the genes and the sequence would be yellow (y), then the two alleles, white (w) and eosin (w^e), then vermillion (v), then miniature (m), and last rudimentary (r). The normal alleles of these mutations would be y^+, w^+, m^+, and r^+. In Muller's system, the name of the mutant is usually what its appearance looks like. Morgan tried to assign inferred functions to the genes, but his system (in which M is rudimentary and R is miniature) was confusing for students to learn. A capital letter for a mutation as in dp^T distinguishes it as being expressed (and therefore dominant in the heterozygote dp^T/dp^+, which would express the truncated wings and disturbed thoracic bristles of the truncate fly). In contrast, the genotype dp^+/dp would have normal ovate wings. Note, also, that the map distances have slightly changed over the years: yellow at 0.0, white at 1.5, vermillion at 33.0, miniature at 36.0, and rudimentary at 55.1. In an incremental model of how science works in genetics, the changes are piecemeal with modifications and changes reflecting the ever-increasing abundance of data.

used the term "factors" initially. Morgan's group, Alfred Henry Sturtevant (1891–1970), Calvin Blackman Bridges (1889–1938), and Hermann Joseph Muller (1890–1967), quickly dominated the formation of classical genetics and the theory of the gene. Morgan characterized genes as sex-limited (later X-linked) or autosomal (not on a sex chromosome). He and Bridges established the relationship between genes and the sex chromosomes X and Y. Morgan's student Sturtevant worked out the first chromosome map for six X-linked genes.

Genes could revert to normal (back mutations or reverse mutations). They could consist of a set of multiple alleles (white, eosin, apricot, cherry, tinged, and buff) as variants of the normal red eye color associated with the first named white eye mutation, later designated as w for white and w^+ for the normal red-eyed condition.[8] Muller showed through his analysis of beaded wings and truncate wings that these genes were chief genes whose extreme or less intense expressions were associated with modifier genes that acted as intensifiers or diminishers. The two genes also responded to differences in temperature. Higher temperatures during development led to more extreme expression of the mutant trait. Muller could combine modifiers and control the temperature and predict the percent of normal, mild, typical, or extreme expression of these genes and their introduced modifiers for any given cross. It was reductionism run into the ground. No vague instability, contamination, or fluctuation of genes existed nor were single genes being shifted about by the environment to produce these outcomes.[9]

In a series of papers, "Mutation," "Variation due to change in the individual gene," and "The gene as the basis of life," Muller threw his career into the gene and its properties. To Johannsen, the gene was a point on a map as dimensionless as a point to a mathematician. To Muller it was some physical or chemical entity that had a length and width (its number and maximum size determined by dividing the smallest interval on the gene map into the actual chromosome map size). Muller engaged in polemic battle with Bateson's, William Castle's (1867–1962), Richard Goldschmidt's (1878–1958), and even his own sponsor Morgan's views to protect the materialistic and science-grounded view of the gene and purge attempts to give it properties that smacked of vitalism, holism, preformation, mysticism, and other philosophic aspects that were beyond experimental test. Muller was pugnacious, but so were Bateson, Castle, Goldschmidt, and other critics with whom Muller debated.

Fortunately for science, it is rarely the personality that prevails for the long-term success of a theory or interpretation. It is the capacity of an idea or finding to withstand the objections and contradictions hurled at it by competitors. It is as if Morgan was Christy Mathewson on the pitcher's mound and Muller was Babe Ruth at the plate, determined to belt a contradiction out of the park.

The Discovery of Position Effect Broadens the Gene Concept

The classical gene began to encounter new attributes. Alfred Sturtevant and Morgan discovered an unusual property associated with the reversions of bar-eyed flies to round-eyed normal flies.[10] Morgan's laboratory technician Sabra Colby Tice found the bar-eyed fly in 1913. It had some unusual properties. It was incompletely dominant. A female heterozygous for this X-linked gene showed a kidney bean eye shape. Homozygous bar females and hemizygous bar males had an identical narrow slit for their eye shape. Also, when keeping the stock of bar-eyed flies, occasional round-eyed males or females with a kidney bean eye shape showed up.

The genotype for the bar-eyed mutant was represented by the letter B. Its normal allele was represented as B^+, and the heterozygous female as B/B^+. After Sturtevant's detailed analysis, bar eyes were interpreted as B^+B^+/Y in a male. Homozygous bar-eyed females were seen as B^+B^+/B^+B^+, and heterozygous females were represented as B^+B^+/B^+. The phenotype of bar-eyed males or females was designated as (B). The frequency was low; about 1 in 1000 showed this reversion from mutant to normal. Morgan and Sturtevant thought this too high. Sturtevant believed it might be related to crossing-over, so they designed an experiment with markers on the left and right of the bar-eyed gene and tested for recombination. They found that the recombination could go either way, with the markers suggesting some sort of mispairings going on within the gene. They suggested it was a duplication of the gene and this made pairing occasionally displaced leading to triplications (an even smaller bar shape) and normal flies (nonduplicated region for that location on the X map). This also meant that bar eyes were not a point mutation.

If there was no lesion in the normal allele for that region, what caused the mutant bar-eyed expression? Morgan and Sturtevant named it a "position effect." It was unexpected and something new about the expression of a gene. A gene could be altered in expression by the genes abutting it. Muller's X-ray technique to induce mutations also induced chromosome rearrangements, and soon rearrangements were found that led to the formation of bar-eyed flies. These, too, turned out not to be alterations of that gene. They were juxtapositions near that gene that could be separated by crossing-over and restored to normalcy again.

Position effect, as Muller's student Carlos Alberto Offermann (1904–1983) showed in 1935, might give three locations for the same gene.[11] It could be a point mutation within that gene; it could be a rearrangement next to the centromere side of the gene; or it could be a rearrangement next to the telomere side of that gene. Suddenly, indeterminacy was introduced into biology. It would not be until the 1960s, when the regulation of gene expression was worked out in

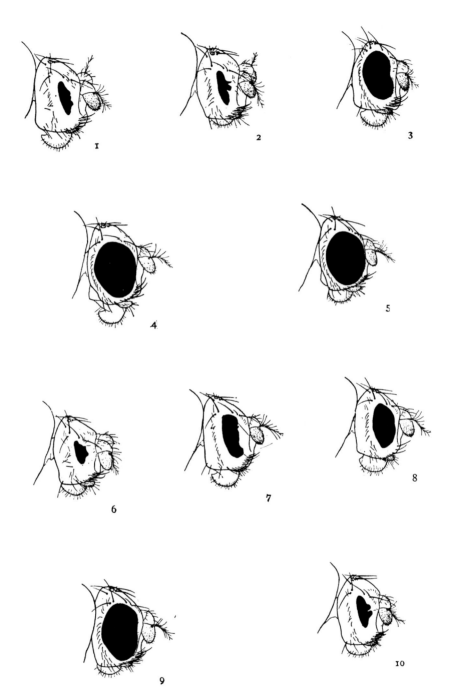

See figure legend on following page.

bacteria as an operon model, that the possible physical and chemical basis of position effect could be resolved.[12]

When the chromosomal region abutting a gene was largely heterochromatic, a new type of position effect occurred. The mutant trait was mosaically expressed from then on. For the white-eye gene, this would be streaks or spots of white against a red eye. Muller called these "ever-sporting displacements." Jack Schultz (1904–1971), who discovered these using X rays, also discovered that an addition of extra heterochromatin (especially an added Y chromosome) would normalize the gene expression and the variegation ceased.[13] This too remained without interpretation until some 30 years later when the molecular basis of gene action and the role of ribosomes in protein synthesis emerged.

Note that breeding analysis was the only effective way classical genetics could develop until 1927, when Muller introduced the artificial induction of mutations with X rays. This tool gave geneticists the opportunity to induce gene mutations in any species and to discover that X rays also induce chromosomal rearrangements through chromosome breaks induced by the radiation. Each time a new technique was introduced, new findings were likely to emerge that could not be anticipated by theory.

Complex Loci, Gene Nests, and Pseudoalleles

In 1940, Clarence Paul Oliver (1898–1991) was studying flies with lozenge eyes. The eyes of lozenge glossy flies had a glassy sheen to them and the eyes of lozenge

---

Sabra Tice in Morgan's laboratory found bar eyes in 1914. It was on the X chromosome, and stocks of bar-eyed flies kept producing round-eyed flies. Sturtevant and Morgan in 1928 showed that bar was a tandem duplication. In the homozygous state, pairing could be between any one of the two duplicate genes and recombination would generate a reversion to normal or an even more extreme ultra-bar (with three tandem bar genes). Both Bridges and Muller in 1936 found bar to be a cytological duplication in the giant chromosomes that were newly discovered by Theophilus S. Painter (1889–1969). Bar was called a position effect mutation. The gene was normal, but shifting it to another place somehow led to its mutant expression. That idea in 1929 eventually became part of the regulatory genetic ideas of operons in the 1950s and epigenetic modifiers in the early 21st century. Sturtevant's 1925 *Genetics* paper describes in detail the various bar eye mutations (shown here) and their capacity to revert to normal. What was novel was his identification of the bar eye phenotype as a position effect not a mutation within the gene. It would take some 50 more years before position effects were recognized as widespread epigenetic events in which genes could be turned on or off or modulated in output. *Drosophila* eye mutations: *1.* Homozygous bar female; *2.* bar male; *3.* bar-over-round female; *4.* female homozygous for round, obtained by reversion; *5.* male that carries round, obtained by reversion; *6.* double-bar male; *7.* homozygous infra-bar female; *8.* infra-bar male; *9.* infra-bar-over-round female; *10.* double-infra-bar male.[21]

spectacle flies had a pigment puddle in the periphery of the eyes. When Oliver crossed these two different alleles and was sorting out the offspring from females that had two such alleles, he found among the expected offspring an unexpected one that seemed to have reverted to the normal. He immediately thought he found a case similar to bar eyes and put appropriate genetic markers on the chromosomes of these two genes and repeated the experiment. Instead of finding the markers shifting in either direction as in the bar-eye case, he found a consistency, with lozenge-spectacle and lozenge-glossy in a fixed order.[14] The term "pseudoallelism" was applied to this new phenomenon. Barbara McClintock (1902–1992), who introduced that term earlier, had found a similar situation in maize, but in her case the allelic difficulties were associated with a deletion in the region of two consecutive genes. In Oliver's case, as Edward B. Lewis (1918–2004) and others would soon show, no deletion was involved in their series of alleles. Lewis thought of the phenomenon as a new type of position effect.[15]

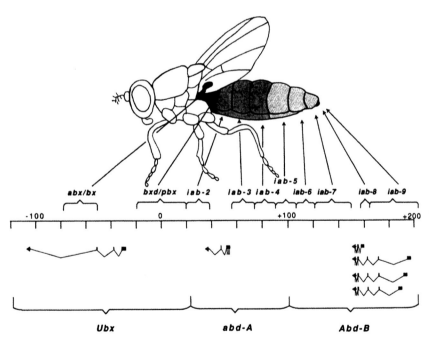

The bithorax series of mutations studied by E.B. Lewis showed how segments normally form the thoracic region, wings, and halteres in fruit flies. The normal fly has two wings and two halteres. Lewis combined genes to produce flies with four wings and other variations in leg or haltere replacement. They were shown at a molecular level to be mutations in a developmental genetic system called the homeobox.

My own study of the dumpy locus (which includes truncate wing) and its alleles showed it to be very complex, with some alleles showing complementarity (apparent nonallelism) as if they were unrelated genes, and yet each of these complementing alleles could show pseudoallelism with other members of the dumpy series. I constructed a complicated map for all the alleles available for testing and even induced several more with X rays and mapped those to show a consistency for where they mapped. Muller used the term "complex loci" for these "gene nests," and he interpreted them as having arisen by unequal crossing-over leading to tandem duplications.

Calvin Bridges, using a cytological approach with salivary gland chromosomes that had enlarged banded chromosomes, also noted such tandem duplications were abundant in the giant chromosomes. He called these "repeats" and attributed the evolution of the chromosomes to such newly arising tandem duplications.[16] In Muller's analysis of the bar-eyed case using salivary chromosomes analyzed by Alexandra Prokofyeva-Belgovskaya (1903–1984), Muller argued that the initial bar eyes found by Sabra Tice arose from a primary unequal crossing-over and that homozygous bar eyes in females generated a secondary unequal crossing-over leading to occasional triplications or reversions to the normal. Muller also offered a "gene doctrine" in parallel to the Remak–Virchow cell doctrine and claimed "all genes arise from preexisting genes," except for the first arising molecule having the property of copying its variations.[17]

Microbial Genetics Generates Genetic Fine Structure

Pseudoallelism soon gave way to a different approach using microbial systems. Viruses and bacteria could yield millions of progeny in a matter of hours or days. Max Delbrück (1906–1981) and Salvador Luria (1912–1991) showed that bacteriophage (bacterial viruses) had a chromosome that could be mapped with spontaneously arising or induced mutations.[18] Joshua Lederberg (1925–2008) showed sexuality in bacteria with the interruption of the sex act (the transmission of the chromosome from one bacterium into another) used to identify the location of the mutant genes being introduced.[19] These studies were pre-DNA, mostly in the late 1930s to late 1940s. The structure of DNA would not be worked out until 1953, and the sequences of nucleotides would not be worked out for DNA until the 1970s when Frederick Sanger (1918–2013) in England and Walter Gilbert (b. 1932) in the United States established techniques for DNA sequencing.[20] But the classical gene would rapidly shift into the molecular gene after 1953.

The theory of hereditary units was born in the 1860s. It picked up experimental support after 1900 and the reintroduction of Mendelism, with breeding analysis its major tool. It would become the theory of the gene and join with the chromosome theory in Wilson's analysis of sex determination. It would fuse with chromosome mapping, nondisjunction, gene character analysis, multiple allelism, and other aspects of classical genetics in the first third of the 20th century. It would become turbid again in the generation between Morgan's "theory of the gene" and the development of microbial genetics, which introduced biochemistry and eventually molecular biology into that theory. During that period of turbidity, concepts of position effect, pseudoallelism, ever-sporting displacements, delayed effects in new mutation formation, dosage compensation, and other phenomena would be primarily descriptive. Classical genetics had

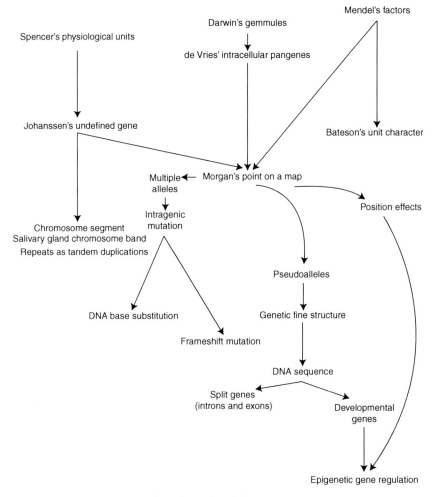

See figure legend on following page.

limited tools from biology (breeding analysis primarily), physics (X-ray induction of mutations and chromosome breaks), and chemistry (the introduction of chemical mutagenesis by Charlotte Auerbach [1899–1994] in the 1940s). It was not paradigm shifts but the influx of tools and findings from microbial genetics, biochemical genetics, and molecular biology that dominated the last half of the 20th century and clarified these classical phenomena.

Split Genes and Epigenetics

When something new is introduced, like the term "gene" in 1909, the easiest aspects are first used or interpreted. The gene was a unit of transmission of a specific trait that could be expressed in Mendelian ratios by appropriate breeding analysis. Those aspects did not indicate where these genes were in the nuclei of cells and did not provide clues to the chemical composition of the gene or how genes function. As genes became assigned to chromosomes and then mapped, the properties of genes emerged. There could be multiple alleles of a given normal gene. There could be epigenetic features like position effect and dosage compensation. The number of genes could be roughly calculated by using

---

In the 1860s, three independent proposals were made for hereditary units in the cell. Spencer called them physiological units. Darwin called them gemmules in a broader theory of pangenesis. Mendel called them factors. In 1895, de Vries changed Darwin's units into intracellular pangenes. Johannsen in 1909 trimmed the term "pangene" into "gene" and left it undefined. It replaced Bateson's unit character after he had read Mendel's paper in 1900. Morgan's fly laboratory (starting in 1911) used crossing-over to map genes as points on a line representing the chromosome on which they reside. They recognized the gene as a small segment of a chromosome and that a gene can mutate to form a series of multiple alleles. Later (in 1940), geneticists found tandem genes that were pseudoallelic. They also found gene expression shifted when a gene was shifted to a new area of a chromosome. They called these placement changes position effects. The discovery of giant cable-like salivary chromosomes in 1936 led to cytological studies of genes. Some, like bar eyes, were associated with duplications of the gene rather than mutations within a normal gene. Most gene mutations resulted in partial or complete loss of function. Such intragenic mutations were mapped by a process producing genetic fine structure. The correspondence of genetic fine structure maps to sequences of nucleotides in DNA led to classification of effects by mutagens, including base substitution mutations and frameshift mutations. The existence of position effects suggested regulatory genes might exist. These were first shown in bacteria in the 1950s, and François Jacob (1920–2013) and Jacques Monod (1910–1976) called them operons. Studies of eukaryotic genes showed a surprising organization of genes as split into functional (exon) and non-functional (intron) sequences. Studies of developmental genes revealed that genes can be turned on or off, or their rates of activity can be regulated by epigenetic factors (including roles for several types of RNA small molecules and for methylation of DNA bases).

relatively well-explored linkage maps. It would take a half-century to shift to biochemical explanations of the composition of genes (nucleic acids, especially DNA for eukaryotic cells). It required a new field, molecular biology, to work out gene structure as a double helix of DNA. It would require new tools to work out the flow of information from DNA to RNA to protein. It would require new technology to sequence genes to their nucleotides. Genes were soon shown to consist of introns and exons. The processes of transcription and translation had to be worked out. Genes could be regulated as operons or regulated by epigenetic factors, especially methylation, or association with transposable elements, especially a variety of RNA molecules. None of these later stages were predictable from the classical phase in the first quarter of the 20th century.

The theory of the gene shifted from one bound largely to description and inferred functions to one bound largely to molecular interpretation. The unit of transmission remained the gene, however dismembered its components became through molecular analysis. Genetic counseling was still primarily Mendelian. The theory of the gene was bound to evolution, medicine, agriculture, pharmacy, and population studies, and it infiltrated every field of the life sciences. Genes were associated with specific traits, they underwent mutation, they were transmitted on chromosomes, they replicated their mutational errors, they occasionally reverted, and, most important, they produced the enzymes for metabolism and structural components of the cell's organelles.

References and Notes

1. Carlson EA. 1966. *The gene: A critical history.* Saunders, Philadelphia.

2. Iltis H. 1932. *Life of Mendel.* Trans. Paul C, Paul E. W.W. Norton, New York.

3. Spencer H. 1851. *Social statics, or the conditions essential to happiness specified, and the first of these developed.* Chapman, London.

4. Darwin C. 1868. *The variation of animals and plants under domestication.* Murray, London.

5. de Vries H. 1889. *Intracellular pangenesis.* Trans. Gager CS. 1910. Open Court Publishing, Chicago.

6. Bateson W. 1902. *Mendel's principles of heredity—A defense.* Cambridge University Press, Cambridge.

7. Johannsen W. 1909. *Elemente der exakten Erblichkeitslehre.* Fischer, Jena, Germany.

8. Morgan TH, Sturtevant AH, Muller HJ, Bridges CB. 1915. *The mechanism of Mendelian inheritance.* Holt, New York.

9. Muller HJ. 1918. Genetic variability, twin hybrids and constant hybrids in a case of balanced lethal factors. *Genetics* **3:** 422–499. Also see Altenburg E, Muller HJ. 1920. The genetic basis of truncate wings—An inconstant and modifiable character in *Drosophila. Genetics* **5:** 1–59.

10. Sturtevant AH, Morgan TH. 1923. Reverse mutation of the bar gene correlated with crossing over. *Science* **57**: 746–747.

11. Offermann CA. 1935. The position effect and its bearing on genetics. *Izv Akad Nauk SSSR VII series* (Bulletin de l'Academie des Science de l'URSS; Russian and English text) 129–152.

12. Jacob F, Monod J. 1961. Genetic regulatory mechanisms in the synthesis of proteins. *J Mol Bio* **3**: 318–356.

13. Schultz J. 1936. Variegation in *Drosophila* and the inert chromosome regions. *Proc Natl Acad Sci* **22**: 27–33.

14. Oliver CP. 1940. A reversion to wild-type associated with crossing over in *Drosophila melanogaster*. *Proc Natl Acad Sci* **26**: 452–454.

15. Lewis EB. 1955. Some aspects of position pseudoallelism. *Amer Nat* **89**: 73–89.

16. Lewis EB. 2003. C.B. Bridges' repeat hypothesis and the nature of the gene. *Genetics* **164**: 427–431.

17. Bridges CB. 1936. The bar "gene" a duplication. *Science* **83**: 210–211. Also, Muller HJ. 1936. Bar duplication. *Science* **83**: 528–530.

18. Delbrück M, Luria S. 1943. Mutations in bacteria from virus sensitivity to virus resistance. *Genetics* **28**: 419–511.

19. Lederberg J. 1947. Gene recombination and linked segregations in *Escherichia coli*. *Genetics* **32**: 505–525.

20. Sanger F, Coulson AR. 1975. A rapid method for determining sequences in DNA by primed synthesis with DNA polymerase. *J Mol Biol* **94**: 441–448. Also, Gilbert W, Maxam A. 1977. A new method for sequencing DNA. *Proc Natl Acad Sci* **74**: 560–564.

21. Sturtevant AH. 1925. The effects of unequal crossing over at the bar locus in *Drosophila*. *Genetics* **10**: 117–147.

CHAPTER 4

Mutation: From Fluctuating Variations to Base Alterations

Sports, atavisms, fluctuating variations, homeotic variations, meristic variations. Mutations as losses, partial losses of genes, reverse mutations, Muller's morphs, induced mutations, mutation by unequal crossing-over, chemically induced mutations. Linearity versus threshold effects, transitions, transversions, intercalating agents, and frameshift mutations. Directed mutations, knockout mutations.

Living things are remarkably diverse. There are millions of species, but we are familiar with only a few hundred by name and a few thousands by type when we encounter them. With the exception of identical twins, no two humans look alike, and even monozygotic twins have differences that their closest relatives and friends can detect. Identical twins never have identical fingerprints. Both random new mutations and environmental influences on gene expression ensure that identical twins are not truly identical. To his surprise, a reindeer herder in Lapland told Carl Linnaeus that he could name and tell apart each reindeer in his flock.[1] Sexologist Alfred Kinsey told me when I was a graduate student at Indiana University that as a boy growing up in New Jersey he could identify each bird in his neighborhood by the variations of the songs they sang. When I see a fruit fly land on my hand, I can tell with my unaided eye that it is a male or a female. In all likelihood you cannot. Familiarity breeds recognition.

The Terminology of Breeders and Hobbyists

In the early 19th century, the Industrial Revolution had produced a leisured and moneyed class who had sufficient income and time to collect things, pursue hobbies, enrich their understanding, and support a burgeoning business of sellers of exotic plants, butterfly collections, varieties of budgerigars, breeds of dogs, new varieties of flowers, and other novelties of the living world. Breeders were guided by the principle of "like for like" heredity. For dogs, that meant careful attention to the desired features for show animals. Those not meeting that standard became "rogues," and either they did not get to breed or they were sold at a lower price and not certified as authentic. At the same time, the new wealth was finding its way to support expeditions around the world bringing in ever-new numbers of new species and varieties from far-away continents and islands. The museums, universities, breeder associations, and wealthy amateurs tried to analyze this daunting avalanche of new varieties. The description of these new organisms was not as challenging a task as the interpretation.

Mutations as Fluctuating Variations, Sports, and Anomalies

In 1830, most scientists believed species were fixed in their basic defining attributes. They also were aware of two kinds of variations. Most were "fluctuating variations," familiar in our fellow humans who differ in very subtle minor variations in every aspect of their appearance.[2] Those fluctuating variations were distributed to the children of parents who could see familial likenesses and at the same time departures from themselves, perhaps some traits going back to a grandparent or an aunt or uncle. There were also dramatic changes that occasionally happened when breeders were plying their trade. A branch of an orange tree may bear fruit that lacked seeds. A sheep might be born with legs so small that no expensive fence was necessary to keep a herd of them within one's grazing property. A pig might be born that as an adult remains the same size as a piglet. Sometimes these sudden appearances were deformities or pathologies of no value to the breeder or collector. In humans, they were often called monstrosities or malformations by the physicians who described them. Generally, scientists called them "sports" or sudden appearances.[3] They were usually "rogues" for the breeders with an eye to showing their birds or animals. They had a more favored future among plant breeders as dramatic shifts in height, branching, floral color, streaking of petals, enriched odors of roses, serration of the leaves, hairiness of the leaf surface, and other features had an appeal to hobbyists.

No one knew why some traits showed fluctuating variations and other traits appeared as sports. In general, the fluctuating variations were compatible with the species' traits. And in general, the sport was a departure that would not survive without human help. Branches with seedless oranges had to be grafted to new trees. It was rarely that clear-cut. Bulldogs were bred for huge jaws and heads and narrow hips. Eventually, the hips were too narrow for females to give birth and the most desired bulldogs had to be delivered by cesarean section. Until the 20th century, variations were considered distinct from heredity. A species' heredity was considered to be held in common for all members of a species. Variations were considered departures.

Darwin's Perception of Mutation Was Shaped by Breeders' Terminology

Charles Darwin was familiar with this issue when he was a young naturalist about to set sail in 1831 on *HMS Beagle*. Five years later, he returned with a wealth of new knowledge, substantial readings of the books he took with him, and a changed attitude toward the "mystery of mysteries," as he referred to the variations within a species and the relation of species to their habitats and environments. Even more radical was his attempt to relate the similarities and differences of species to one another. In 1838, the subversive idea came to him that he could interpret this change through a theory of evolution by natural selection. He did not publish this, but he expanded his ideas in 1842 and 1844 in two larger unpublished essays that gave the broad outlines of what in 1859 would become *On the Origin of Species.*[4] Darwin had shifted mutation, the study of variations, from practical breeders to scientists. He also shifted the study of species from those who like to classify them, who like to display them in zoos, or who like to exploit them for their commercial value. Instead, that biological study of species now became central to the theory of evolution he promoted. Finally, he shifted the study of species from an ancient religious belief of the fixity of species through acts of a Creator to a natural process that required no guiding hand of a Creator or a magical introduction of species either in the seven days of Genesis or in separate acts of divine creation imagined by geologists studying the past history of life buried and fossilized in rocks.

Bateson Favored Discontinuous Variation for Evolution

Darwin favored fluctuating variations as the material on which natural selection acted. Sports were too radical a departure from normalcy and too pathological in their effects on organ systems to breed on their own and in competition with the environment and habitat they encountered. For the rest of the 19th century, Darwin's views prevailed as naturalists sought to find ways to measure the evolutionary changes of fluctuating variations in natural populations. This was a statistical approach to evolution that led to a flourishing school of biometrics headed by Francis Galton (1822–1911) and his protégés, Karl Pearson (1857–1936) and W.F.R. Weldon (1860–1906).[5]

Against that solid core of Darwinians came an upstart, William Bateson, who became radicalized during his sojourn at Johns Hopkins University, and by the somewhat philosophic musings of William Keith Brooks (1848–1908), who told Bateson never to forget that the problems of heredity should be solved by science and that theory could not go farther to explain it. Bateson came back with the belief that variations need not be continuous and fluctuating to bring about species. He spent his weeks in the library gathering a massive collection of sudden mutations, most of them sports, and he reclassified them. Some were repetitions of parts like extra digits in a human or cat. Some were like extra vertebrae in a mammal. He called these repetitions of parts "meristic variations." Others showed the shift of one organ system to a new location, like a tuft of feathers on a chicken's head instead of a fleshy comb. On an insect, what should be an antenna might instead sport a leg. Bateson called these "homeotic variations." He argued that these meristic and homeotic variations, usually appearing as isolated sports, along with changes in body symmetry, could account for the more dramatic changes in species. He published his compendium in 1894 and attacked the biometric school for not facing up to the beneficial effects such meristic and homeotic changes could bring about.[6]

The biometricians ridiculed Bateson. When Mendelism showed how discontinuous traits could be distributed in fixed laws, Bateson seized on the new findings and rapidly extended Mendelism, finding new phenomena such as epistasis (gene interaction giving modified Mendelian ratios), recombination (which he called repulsion and coupling), and non-Mendelian ratios associated with recombination (which he attributed to a differential reduplication of the alleles present in the gonadal cells). In 1906, he named the new field "genetics," paying tribute to Darwin by preserving the root "gen" in his theory

Meristic mutation of deer antlers. Note the mosaicism in both specimens, suggesting an origin by mutation in a blastomere rather than in a reproductive cell. Bateson identified these specimens as follows: "Abnormal horns of Roebuck (*Capreolus caprea*), No. 438. (When seen by me the horns were fixed upon heads modelled in plaster.)"

of pangenesis. The more Bateson published, the more Pearson and Weldon reacted with bitterness. Galton tried to play to both sides and set up a committee in the Royal Society to explore evolution and heredity and put both Bateson and Weldon on it. The meetings were lacerating and Weldon quit and died shortly afterward. Bateson used the "Reports to the Evolution Committee" as a forum for his publications that were otherwise blocked by Pearson.[7]

Morgan's Fly Lab Encounters New Mutations and Their Properties

The shift in thinking about mutation came about from T.H. Morgan's laboratory. The mutations were discontinuous and could be mapped in a line reflecting the inferred string of genes linked along the length of a chromosome. The mutations were not just losses as Bateson claimed, because the allele called eosin arose in a stock of white-eyed flies. Morgan dropped the term "variation" and used the term "mutation" because he was impressed by de Vries' work on the evening primrose, *Oenothera lamarckiana*. Morgan had visited de Vries in Holland and returned with enthusiasm to Columbia to embark on an extension of de Vries' "mutation theory."[8] In de Vries' theory, new species arose suddenly, just as new species of primroses had appeared in his garden. They could breed with themselves but not with their parental type. They differed in numerous characteristics from that parental type. Morgan switched to fruit flies in 1905 on the advice of William Castle, who had used them for variations in wing size and venation to see if Darwinian fluctuations could be confirmed in these fly cultures. Morgan saw the opportunity to find new species in these flies. After two years of effort, he got discouraged—he had found nothing. In 1909, things began to change. He found a fly that showed a streak on its thorax. Later, he found a fly with a dark pigmentation at the juncture of the wing to the thorax. His third mutation, in 1910, gave him something clear-cut, a white-eyed male. The white-eyed fly led to X-linked inheritance. More X-linked traits led Morgan to the concepts of crossing-over and mapping.[9] Morgan quickly abandoned the hunt for new species and shifted his attention to interpreting the new mutations that cropped up. He expanded his laboratory and was soon surrounded by eager graduate students hoping to share in the excitement Morgan felt with each day's adventures in the fly lab.

Why the Fly Lab Won in Their Battles with the Biometricians, Bateson, and de Vries

Curiously, Bateson was not enthusiastic about Morgan's findings. Morgan and his students were connecting mutations to the chromosome theory. Morgan had abandoned evolution for a narrow genetic aspect of the broader picture Bateson hoped to find. Similarly, de Vries was not happy about the way Morgan's enthusiasm had waned; Morgan was mired in chromosomes instead of speciation. All three had a common foe in the biometric school, but instead of presenting a united front they offered three different approaches to mutation. Bateson sought mutations that could lead to evolutionary change by meristic and homeotic mutations. He did not believe new species arose de novo with multiple differences from their parent. Morgan, who almost took pleasure in abandoning his theories whenever a new experiment revealed something new, overcame his own skepticism and now became a zealous advocate of the chromosome theory of heredity. de Vries kept waiting for a "mutating period" to erupt in Morgan's work or Bateson's work. Neither obliged him. He saw his carefully crafted mutation theory of evolution dismantled piece by piece as his new species turned out to be polyploids or complex chromosomal rearrangements that other geneticists quickly showed. *Oenothera* was not the leader of a new movement and was shifted into the rear as an oddity of speciation found in virtually no other species.

For the biometricians, the squabble among their opponents was of little comfort. A statistical analysis of claw length or ratios of limbs to body length revealed little about the evolutionary process. Thus, their journal articles repeated the obvious—there were quantitative and subtle differences in species and varieties that could be described in bell-shaped curves and more complex ways, but they told nothing about the biological mechanisms of evolution. By 1921, the work of Morgan's group was showing how chief genes and modifiers could result in predictable categories of offspring. Those genes could be mapped. They could be recombined in experiments and tested for their influence on character traits.[10] Discontinuous mutations were producing continuous variations in expression. These results were not just being produced in the fly lab. Similar assaults were coming from studies of quantitative traits in cereal grains, which showed some strains with three or four pairs of genes producing bar graphs of coloration differences similar to bell curves.[11] Wilhelm Johannsen could distinguish genotype from phenotype through selection experiments involving bean size.[12] Classical genetics was now leading the way to a union with Darwinian evolution. By the 1940s, this would be called "the new synthesis," with cytology, population genetics, systematics, paleontology, and Mendelian genetics intermingled in their approaches to evolution. It was thoroughly Darwinian in its support, but vague fluctuating units of inheritance had been rendered precise through the chromosome theory of heredity that Muller likened to a dissection of heredity at the cellular level.

The Induction of Mutations by X Rays Provides a New Tool for Genetics

In 1927, mutation took on new significance when H.J. Muller reported the induction of gene mutations with X rays.[13] He obtained more mutations in one winter's work than all fruit fly geneticists throughout the world had reported since Morgan reported his first mutants in 1909. He also found new phenomena. Some of these mutations were repressors of crossing-over, which he discovered when trying to map them. They turned out to be chromosome rearrangements such as translocations and inversions. They were caused by chromosome breakage followed by the reunion of pieces in these novel ways. Also found in abundance were a curious group of dead embryos laid by the females who received irradiated semen. It would not be until the 1940s that these were shown to be chromosome breaks leading to dicentric chromosomes that aborted the embryos because they could not carry out a normal mitosis. Radiation genetics emerged as a specialty field of genetics because gene mutations were induced in linear proportion to the dose received. But inversions or translocations were induced with a nonlinear increase that was approximately to the 3/2 power. The theoretical square of the dose difference did not occur because in a mature sperm that is irradiated the chromosomes are compressed, coiled, and packed into the sperm head, which is virtually void of protoplasm. Many ionizing paths would cut through two or more chromosomes and this would diminish the expected square for independent breaks in the chromosomes by separate paths of ionization.

Genetics became complicated. Its findings became comprehensible to others in the field, but to other biologists it was something they could not enter. After the atomic bombings of Hiroshima and Nagasaki, Muller interpreted radiation sickness as a consequence, not of gene mutations, but of dicentric chromosome breakage, whose analysis he worked out with Guido Pontecorvo (1907–1999) in 1940 to explain dominant lethals.[14] For the public, mutation was something scary. The public does not discern gene mutations from chromosome breakage. The public does not discern low-dose effects from high-dose effects. Mutation in the public mind is equated with monstrosities and the sports in Bateson's 1894 book filled with the homeotic and meristic variations of nature—two-headed calves, children with lobsterlike claws for hands and feet, children who looked like they were melting candles with their skin hanging in flabs, cats with seven toes, or conjoined twins.

The Genetic Analysis of Gene Function

Muller introduced two interpretations of gene function before a biochemical basis was known for their function. The first is often described by historians

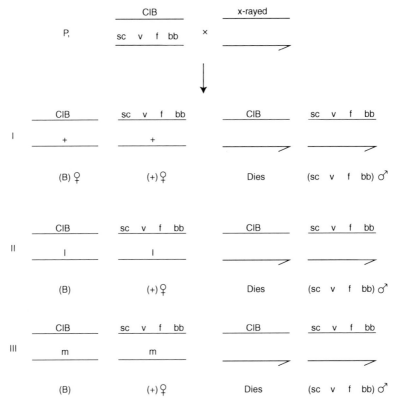

In Muller's experiments presented in Berlin 1927, he described the ClB stock he devised to detect induced gene mutations on the X chromosome. He constructed the ClB chromosome, which had a crossover suppressor C, a recessive lethal l, and a dominant mutation bar eyes B. The females with the ClB were heterozygous for an X chromosome containing the marker genes scute bristles (sc), vermilion eye color (v), forked bristles (f), and bobbed bristles (bb). He then exposed males that had a normal X and Y chromosome to X rays. Three types of results emerged. In category I, the vials contained bar-eyed females, normal-looking females, and males that were sc v f bb. In category II, the vials contained bar-eyed females, normal-looking females, and males that were sc v f bb. In category III, the vials contained bar-eyed females, normal-looking females, and males that were sc v f bb. Muller then tested heterozygous B females from each vial by crossing them with their brothers who carried the markers sc v f bb. Examination of the offspring from Category I showed one category of males all with normal eyes. Category II showed no males at all, indicating the female carried the induced lethal. Category III showed one class of males carrying a newly induced visible mutation from the X-ray exposure. From this, Muller could calculate the induction of mutations at any given dose. Most of the induced mutations on the exposed male X chromosomes were recessive lethals, killing the embryo after fertilization and before emerging as adults. The normal-looking females of those in category II had the induced lethal, which could be mapped among the genes sc v f bb in those females. Similarly, those in Category III could have their new visible X-linked mutation mapped among those marker genes.

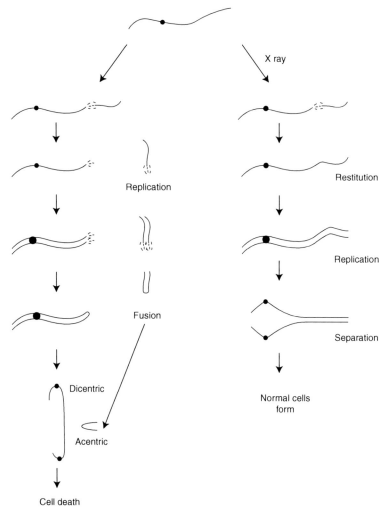

In the breakage–fusion–bridge cycle, Barbara McClintock in Missouri and independently H.J. Muller and Guido Pontecorvo in Scotland worked out the breakage–fusion–bridge cycle. In this process, a chromosome experiences a spontaneous induced chromosome breakage. On the *right*, the break is repaired by an enzyme, restituting it. It replicates, separates the two chromatids, and normal cells result. On the *left*, the break occurs and the fragments separate, drifting apart, and cannot be repaired. After replication, the adjacent broken ends of the centric fragment join. So, too, do the replicated broken ends of the acentric fragment. During cell division, the acentric chromosome cannot be attached, and those genes are lost to the cell. The centric fragment forms a chromosome bridge preventing the cell from dividing, and it eventually dies. Muller in 1945 interpreted the symptoms of radiation sickness as a consequence of breakage–fusion–bridge cycles in the dividing tissues of the people exposed to atomic bomb radiation in Japan.

of science as "Muller's morphs." He noted that in the white-eyed series of alleles, the colors ranged from white to near-red. Most were white. The nonwhite alleles were given names like ecru, buff, or eosin, suggesting a pale pink color. Collectively these off-white colors he described as hypomorphs. They had lost some, but not all, of their capacity for making pigment. The original white-eyed fly that Morgan found in 1910 produced no pigment, and Muller called that condition an amorph. An amorph could be a total loss of the gene (by deletion) or a total inactivation of the gene. Mutations that produced excess pigment, such as ebony body color, he called hypermorphs.

The second function Muller called dosage compensation. It was a puzzle to Morgan and his students that males with one X chromosome usually produced the same phenotype (e.g., eye color, wing shape, bristle morphology) as females with two X chromosomes. There were some exceptions, one being the white-eyed allele, eosin, which was darker in females (XX) than in males (XY). Why were most X-linked genes dosage compensated and what was that mechanism? Muller designed experiments using deleted X chromosomes that contained the region for the white-eyed alleles and identified genetic modifiers on the X chromosome involved in the dosage compensation process. He presented his experimental analysis of gene function (the morphs) and dosage compensation in 1932 at the International Congress of Genetics in Ithaca, New York.[15]

Induced Mutations Become the Tool for Determining Biochemical Pathways

Mutation shifted again with the introduction of biochemistry and molecular biology starting in the late 1930s. George Beadle (1903–1989) used X rays to produce mutations in the orange bread mold *Neurospora*. What he sought were mutations that prevented the cell from making a vitamin. He isolated such mutations and showed that they prevented the formation of several different enzymes that he and his colleagues could isolate and plot into a biochemical pathway from the smallest initial precursor molecule to the final product in this chain of construction. Each enzyme added to the molecule or tweaked it to alter its shape. Mutations were now recognized as valuable tools for studying metabolism through this approach. More important for Beadle was the implication that each gene seemed to be involved in the production of a specific protein or enzyme involved in the cell's living functions. He called this the one gene–one enzyme theory.[16] This led to the union of biochemistry and genetics and also promoted the use of microorganisms for the study of biochemical or molecular aspects of mutation. What Beadle also introduced was a shift in team research from two specialists in the same field to two specialists in cognate fields.

He had teamed with Edward Tatum (1909–1975), a biochemist, to work out the pathways from the gene mutations Beadle isolated.

Mutation Shifts to the Study of DNA

In 1945, DNA was identified as the genetic material in studies of the bacterium that causes pneumonia, *Diplococcus pneumoniae*. A team of scientists at Rockefeller University headed by Oswald Avery (1877–1955) showed that isolated DNA (but not RNA, protein, or carbohydrates) extracted from virulent strains could be used to alter nonvirulent strains of that bacterium. If this "transforming principle," as they called it, was digested with an enzyme, it failed to carry out transformation.[17] In 1953, the structure of DNA was shown to be a double helix of aperiodic sequences of nucleotides held by hydrogen bonding associated with complementary pairs of nitrogenous bases. The biological implications were stated by J.D. Watson (b. 1928) and F.H.C. Crick (1916–2004) in a follow-up note to *Nature*.[18] They claimed their model would explain gene (or chromosome) replication, the nature of mutations, and the way genes work. Replication was quickly associated with the pairing relations of adenine (A) with thymine (T) and of guanine (G) with cytosine (C).

The tools and information used by Watson and Crick included a number of approaches. For theory, they relied on Erwin Schrödinger's book, *What is Life?* Schrödinger argued that the gene was an unusual crystal because it did not have repeating components along its length. The gene was aperiodic.[19] It had the properties that Muller had identified in the early 1920s as "covariant reproduction." The gene could replicate its errors or mutations. Watson, who had taken Muller's course on "Mutation and the Gene," was familiar with that thesis. They also used X-ray diffraction. Here Crick played the major role because he could interpret the X-ray images (produced in Maurice Wilkins' laboratory by Rosalind Franklin and Raymond Gosling) that revealed the crystalline structure of DNA. Their third approach used something that could be called "model building." Watson played a major role in that. He prepared cardboard cutouts of the nitrogenous bases and, by moving them around, he was able to show a fixed pairing relation of A with T and of G with C. This explained the Chargaff ratios in which A + G (purine) = T + C (pyrimidine) and A + C = G + T (matching hydrogen bond pairing sites), but A + T is not equal to G + C (the basis for an aperiodic sequence or genetic code). Their fourth approach was chemical. They (mostly Crick) used chemical texts to get the exact shapes and angles of the atomic components of the nitrogenous bases to construct the three-dimensional model of their double helix. Their fifth approach was biological. The base sequence had to be aperiodic for DNA to have the properties of a gene.[20]

The recognition of DNA as genetic material did not immediately lead to an analysis of that molecule for the production of mutations. A somewhat different approach was used by microbial geneticists using bacteria or bacterial viruses. Chemical mutagens were chosen rather than X rays for several reasons. Chemicals could be characterized by the type of reactions they had with other chemical substances. X rays produced both chromosome breaks and gene mutations. For the breaks, the results seemed a consequence of a direct hit (or absorption of energy or path of dense ionizations) in the chromosome. For gene mutations, there could be direct hits to the gene or the conversion of nearby water into hydrogen peroxide, which would act as a chemical mutagen. Thus, at least three things could be going on and this made X rays messy for those seeking direct cause-and-effect associations with the mutation process.

Chemical mutagens were first successfully studied by Charlotte Auerbach at the University of Edinburgh. She was a postdoctoral student in Muller's laboratory. He suggested to her that a search for chemical mutagens would be a worthy task for studying the gene. She tried coal tars, but these were mostly insoluble in water and difficult to get into the reproductive cells of flies. She then asked John M. Robson (1900–1982), a pharmacologist, if he could suggest a compound. He suggested mustard gas because it produced blisters and damage similar to that of radiation during World War I when it was

1. ACG.TTC.GA*G*.AAA.GTG---   ACG.TTC.GAA.AAA.GTG---   Transition mutation

2. ACG.TTC.GA*G*.AAA.GTG---   ACG.TTC.GA*C*.AAA.GTG---   Transversion mutation

3. ACG.TTC.GA*G*.AAA.GTG---   ACG.TTC.GA*G.T*AA.AGT.G---   Frameshift mutation

Seymour Benzer's laboratory used chemical mutagens and showed that certain agents like 5-bromouracil or ethyl methanesulfonate (EMS) produced base substitutions if the change was from a purine to a purine or a pyrimidine to a pyrimidine; they called this a transition mutation. *Line 1* shows this as the third triplet GAG, which has substituted an A for a G (a purine for a purine). *Line 2* shows a substitution of a C for a G, or what was called a transversion replacement. Both transitions and transversions are base substitutions. In *line 3*, there is an addition of one or two bases in the DNA reacting with a chemical called an intercalating agent, such as proflavine. It inserts itself in the DNA and distorts the chain of nucleotides when DNA synthesis occurs, causing these micro-errors. In *line 3*, there is an insertion of an extra T between triplets 3 and 4 of the DNA. This results in the reading being frameshifted as it is read to the full length of the gene. Before genes were sequenced as DNA, mutations without known alterations of the chromosomes were called point mutations. Molecular biology classified these into three structural categories: transitions, transversions, and frameshift mutations.

used as a weapon. She and Robson did those experiments, Auerbach exposing the flies (and her unprotected hands) to mustard gas. She got what she described as "heaps of mutations."[21] They had some unusual features. Many of the visible mutations first appeared as fractional or mosaic flies with only one side of the fly affected. If the mosaic individual was bred, only ~20% transmitted that mutant condition to the offspring and that was as a complete mutation. Also, there was a paucity of chromosome rearrangements, which suggested that mustard gas produced gene mutations by direct contact with the gene and that it did not break the chromosome in the process of generating mutations.

The Double Helix and Its Aperiodic Sequence of Nucleotides Changes Mutation Studies

For this reason, chemical mutagens were used on bacteriophage by Seymour Benzer's (1921–2007) laboratory (with Ernst Freese [1925–1990]) and Crick's laboratory (with Sydney Brenner [b. 1927]). They showed that some chemical agents such as ethyl methanesulfonate (EMS) or nitrous acid or hydrogen peroxide produced mutations they characterized as "transitions" and "transversions."[22] The transitions were AT to GC (or GC to AT) changes as the altered DNA replicated. The transversions were AT to TA (or GC to CG) changes. This meant for transitions that one purine replaced another purine in a strand (or one pyrimidine replaced another pyrimidine). In transversions, there was a switching of base type—a purine would be replaced by a pyrimidine (or the reverse). For chemical agents like proflavine or quinacrine, which were multiring structures, the results were different. Most of their mutations seemed to knock out or add one or more base pairs from the gene. This resulted in what Crick and Brenner described as "frameshift mutations."[23] They used the analogy of a keyboard—if one's fingers were misaligned while touch-typing, the typed message came out gibberish.

Whereas transitions, transversions, and frameshift mutations were inferred lesions in the DNA molecule in the 1980s, by the end of the century the DNA itself could be sequenced and the mutational lesions characterized. These confirmed the suppositions made by these two laboratories. The molecularization of genetics by that time also allowed synthetic genes to be made. Any gene could have any base pair replaced, and the effects of this mutation could be tested when it was returned as a replacement in a virus or cell. Today, mutations can be used to identify evolutionary changes by comparing them in known molecules (such as hemoglobin) or in the non-protein-forming so-called "junk" DNA called introns in split genes.[24] Noncoding DNA/RNA has also become an active research area; it carries information and regulates coding regions of DNA. Sequencing also permits leftovers from the evolutionary past. DNA can be extracted from mummified tissues in museum specimens or those retrieved by archaeologists from caves.

Still No Paradigm Shifts for the Idea of Mutations

From this analysis of the idea of mutation, the term has shifted from fluctuating variations, rogues, and sports in the 1830s to the Mendelian factors (later gene mutations) of classical genetics. They shifted as gene interactions that were noted by Bateson for epistasis and by the fly lab's studies of chief genes and modifiers

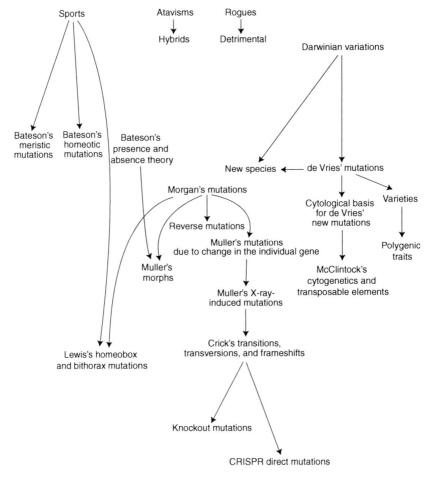

See figure legend on following page.

resulting in gene/character relations. They could be induced by ionizing radiation and other physical agents as well as by chemicals. As genetics went biochemical and molecular, the mutation process could be refined and described by inferred, and later by confirmed, changes in bases or base pairs in DNA. This is incrementalism at work. The story is not a logical pathway of working out a puzzle whose pattern as a whole was apparent to Darwin (or anyone else) as he began his five-year journey around the world. Nor is it a story of random findings later organized by a perceptive paradigm-shifter. It is a story of experimentation, new technologies, a discarding of misleading or false conclusions, and a steady amassing of evidence that mutations are changes in the individual gene and separate from other phenomena such as chromosome breakage and ploidy changes in the cell nucleus. Finally, the mutation process could be seen (but not predicted in its specificity) as a molecular process in which nucleotides are altered, lost, or duplicated within the gene to generate new mutations.

In Darwin's era, breeders talked of "sports" as sudden appearances of a trait. They talked of atavisms as "throwbacks" to an ancestral form. They went through their stocks and discarded (or ate) the rogues that were not true to type. Darwin accepted these terms but focused on what he called fluctuating variations that occurred in nature when he considered the changes selected for evolutionary success. The term "sport" was dropped by Bateson, who used the terms "meristic" (duplication of parts) and "homeotic" (displacement of organs) for these variations. Atavisms were often seen during hybridization between two varieties and the term "hybrid" was used to replace it. Rogues became detrimental mutations. Hugo de Vries in 1900 rediscovered Mendelism. He found in his breeding of evening primroses (*Oenothera lamarckiana*) new varieties and occasionally new species of *Oenothera*. He called these occurrences a "mutating period," and he proposed evolution by saltations or jumps involving numerous characters changing at once to produce new species. Bateson thought of most mutations as a presence or absence of a genetic factor. Morgan's mutations turned out to be simple changes in a gene resulting in partial or complete loss of a trait (like red vs. white eyes). He and his students also found reverse mutations. de Vries' mutations turned out to be changes in polyploidy (triploid, tetraploid) and changes in ploidy number (aneuploidy), usually a gain or loss of a single chromosome and complex rearrangements of chromosome parts. This led to Barbara McClintock's founding of cytogenetics. Muller argued for restricting the term "mutation" to a change in the individual gene. He proposed a functional consequence that he designated as amorphs, hypomorphs, hypermorphs, and neomorphs. Geneticists found some traits involved chief genes and modifiers, others involved quantitative traits, and still others were like Darwin's fluctuating variations. Muller induced mutations with X rays. His postdoctoral student Charlotte Auerbach induced mutations with chemicals. When DNA was identified as the genetic material, Crick identified transitions, transversions, and frameshift changes in the DNA base sequence. Darwin's fluctuating variations led to population genetics. With the molecularization of the gene came knockout mutations in which individual genes could be selectively made nonfunctional. Also, a molecular genome-editing tool, CRISPR/Cas9, has made it possible to locate and replace a gene allowing directed mutation with the introduction of the replacing gene.

References and Notes

1. Blunt W. 2001. *Linnaeus: The compleat naturalist*. Lincoln, London.

2. At issue was whether these fluctuations were of a hereditary nature or minor responses to the environment. Some breeders believed they had no hereditary role and it was sudden mutations or sports that led to new characteristics in a breed. Some believed these new mutations were a consequence of hybridization. Darwin believed the fluctuating variations were the raw material for evolution by natural selection. The term "mutation" was introduced about 1902 by de Vries in his publications on "the mutation theory."

3. The term "sport" was applied by practical breeders to the germline by using the term "seed sports" and to the somatic line by using the term "bud sports." Bud sports sometimes required grafting for propagation.

4. Darwin F, ed. 1909. *Charles Darwin and the foundation of the origin of species, two sketches written in 1842 and 1844*. Cambridge University Press, Cambridge. Darwin C. 1859. *On the origin of species by means of natural selection or the preservation of favoured races*. Murray, London.

5. Cock AG, Fordyce DR. 2008. *Treasure your exceptions: The life and science of William Bateson*. Springer, New York.

6. Bateson W. 1894. *Materials for the study of evolution*. Cambridge University Press, Cambridge.

7. Bateson W. 1910. Experimental studies in the physiology of heredity. In *Reports to the Evolution Committee of the Royal Society I–IV, 1902–1909*. Royal Society, London.

8. de Vries H. 1902. *The mutation theory*. Trans. Gager CS. Open Court Publishing, Chicago.

9. Allen G. 1978. *Thomas Hunt Morgan: The man and his science*. Also see Carlson E. 1966. *The gene: A critical history*. Saunders, Philadelphia.

10. Muller HJ. 1917. An *Oenothera*-like case in *Drosophila*. *Proc Natl Acad Sci* **3**: 619–626.

11. Nilsson-Ehle H. 1909. "Kreuzunguntersuchungen an Häfer und Weizen." ["Crossbreeding studies of oats and wheat."] Academic Dissertation, University of Lund, Sweden.

12. Johannsen W. 1909. *Elemente der exakten Erblichkeitslehre*. [*Elements of an exact theory of heredity*.] Fischer, Jena, Germany.

13. Muller HJ. 1927. The problem of genic modification. Verhandlungen V international Kongress Vererbungslehre. *Z Abstam Vererbungslehre*, Supplement I, 234–260.

14. Pontecorvo G. 1942. The problem of dominant lethals. *J Gen* **43**: 295–300; McClintock B. 1938. The fusion of broken ends of sister half-chromatids following chromosome breakage at meiotic anaphase. *Univ Missouri Coll Agric Res Bull* **290**: 1–48.

15. Muller HJ. 1932. Further studies on the nature and causes of gene mutations. *Proc Sixth Intl Cong Gen, Ithaca NY* **2**: 215–255.

16. Beadle GW, Tatum EL. 1941. Genetic control of biochemical reactions in *Neurospora*. *Proc Natl Acad Sci* **27**: 499–506.

17. Avery OT, MacLeod CM, McCarty M. 1944. Studies on the chemical nature of the substance inducing transformation of *Pneumococcal* types: Induction of transformation by deoxyribonuclear fraction isolated from *Pneumococcus* type III. *J Exp Med* **79**: 137–158.

18. Watson JD, Crick FHC. 1953. Genetical implications of the structure of deoxyribonucleic acid. *Nature* **171**: 964–967. The DNA structure paper is Watson JD, Crick FHC. 1953. A structure for deoxyribose nucleic acid. *Nature* **171**: 737–738.

19. Schrödinger E. 1946. *What is life?* Cambridge University Press, Cambridge.

20. Watson JD. 1968. *The double helix.* Atheneum, New York.

21. Auerbach C, Robson J. 1946. Chemical production of mutations. *Nature* **157**: 302.

22. Benzer S, Freese E. 1958. Induction of specific mutations with 5-bromouracil. *Proc Natl Acad Sci* **44**: 112–119.

23. Crick FHC, Barnett L, Brenner S, Watts-Tobin S. 1961. General nature of the genetic code for proteins. *Nature* **192**: 1227–1232.

24. Roberts R. 1976. Restriction endonucleases. *CRC Crit Rev Biochem* **4**: 123–164. Also, Berk AJ, Sharp PA. 1978. Structure of the adenovirus 2 early mRNAs. *Cell* **14**: 695–711.

CHAPTER 5

The Life Cycle: From Spontaneous Origin to Simple and Complex Stages

Seven stages of human life, spontaneous generation, life from preceding life (Redi). Microbes as organisms (Koch and Pasteur), microbes arise from preceding microbes (Pasteur). Microscopy, neotonous versus imago births, alternations of generations, diploid versus haploid life stages, aging, death.

It seems obvious today that living things have a life cycle. Life cycles were virtually unknown for almost all living things except for larger animals and plants. Until 1864 and the work of Louis Pasteur (1822–1895), there was still a sympathy that life could arise from nonlife. Before 1668 and the experiments of Francesco Redi (1626–1697), most of life was thought to lack a life cycle.

If we think of our own life cycle, how would we describe it at the time of the founding of the United States of America, let us say in Philadelphia July 4, 1776? For one thing, very little was known about embryology and no studies of human embryology had been performed. There was no cell theory. Instead, we would acknowledge a period of pregnancy (about 9 months measured by the cessation of the menstrual cycle in the early stage of pregnancy and terminating in birth). The baby would remain an infant until it became a toddler. It would then go through childhood until adolescence. Adolescence would end as menstrual periods appeared in the young female and a beard in the young male began a tentative growth. Somewhere between 13 and 21 years in different cultures, the person would be considered a young adult. Some women, including American women in the new republic of 13 states, could marry legally as early as their 13th year. Young adults would soon become mature adults until they entered their middle age (about age 30–35 for most adults).

They would be considered old somewhere between 50 and 65. They would show signs of senescence after age 65 with a biblical life expectancy anticipated, for most persons who survived childhood, of about 70 years. Calculating mean life expectancy is more difficult because, until the germ theory was introduced in the 1880s, about half of all babies born did not make it to their second birthday. That depressed mean life expectancy to about 45 years for infants born in the first generation of the new republic. It slowly crept up as conditions improved to about 55 in the 1890s. The public health measures that emerged after the Civil War and the regulation of food and milk by pasteurization and other hygienic measures dramatically shifted survival so life expectancy was about 65 during the Great Depression. By the start of the 21st century, it was closer to 82 for females and 78 for males. It will continue to creep up as nutrition and public health and health awareness are made universal. The biblical expectation of 70 years as our likely life expectancy is now replaced by actuaries determining health insurance and life insurance policies as about 85 for females and 80 for males in the United States. Life expectancy varies a lot around the world. Some countries (Norway, Japan) fare better and some (Uganda, Haiti, Russia) are more like life in the start of the 20th century or even earlier.[1]

Note that today we would recognize fertilization (not worked out until the 1870s) and embryo formation (organogenesis), which involves the first 55 days of a human pregnancy. We would also distinguish the embryo from the fetus. The fetus grows and enlarges and matures its organ systems, but those organ systems were all laid down during organogenesis. In the world before the mid-19th century, most pregnant women would describe "quickening" as an event when they first felt movement of the baby (a fetus). In the Middle Ages, that was believed to be the moment when a soul entered, animating the baby. In medieval theology, abortion was a sin but not a crime of murder before quickening. This was also true for Muslim beliefs.[2]

Changing Views of the Human Life Cycle

Many people over the past two or three millennia recognized only the human stages that could be seen as our life cycle: birth, infancy, childhood, adolescence, adult, middle-aged, old, and finally death to put a closure on that life cycle. That's seven stages. Now it is fertilization, organogenesis, fetal development, birth, infancy, childhood, adolescence, young adult, middle-aged adult, old, senescent, and finally death. We now distinguish 12 stages in the human life cycle. The refinement comes from two sources. The first is the introduction of effective microscopes and stain technology in the mid-19th century. The second is the convincing demonstration that is sometimes given the name "the biogenetic law" or "all life is derived from preexisting life."

The belief that there is a spontaneous creation of life goes back to biblical times. It was compatible with the view that God created life, because fundamentalism did not emerge as a movement until the early 20th century. The idea that living things came from filth, slime, and putrid matter was almost universally accepted. There was no sense that sexuality existed in insects, worms, eels, and many other organisms. Aristotle distinguished between two forms of life. Those

that required sex to breed he called univocal life. Those that arose from spontaneous generation he called equivocal life. He and his contemporaries also
accepted a vital heat or "pneuma" as characteristic of life. Very likely the cooling
of the body when it is dead suggested such a vital heat. Pliny added the possibility
of cloning and suggested eels reproduced by the scrapings from their bodies,
which enlarged and developed into new eels.[3]

Challenges to the Idea of Spontaneous Creation Begin with Francesco Redi

No major challenges appeared for this belief until the 17th century. Francesco
Redi was an Italian physician who got his M.D. in Pisa. In 1668, he showed
that putrefaction was not the source of maggots on spoiled meat. He placed fresh
meat in jars and covered one set with fine cloth (he referred to it as a veiled fabric). A second batch he left open. In both cases, the meat rotted and the ones
lacking cloth coverings were swarming with maggots. He isolated the maggots
and placed them in another jar, which he covered. They developed into familiar
flies. Redi's work established the first insight into the life cycle of flies. They have
a stage that is distinct from that of the adult. When the cloth was placed directly
on the meat and then allowed to rot, Redi observed that the maggots were on the
surface of the cloth but not on the meat itself. This showed him that the flies
were depositing eggs or something equivalent and small that developed into
maggots, which in turn developed into adult flies. He posited that animals are
"more likely born from eggs laid by their mother." From this Redi concluded
that spontaneous generation was false.[4] His critics did not agree. They shifted
spontaneous generation to a microscopic level. They claimed putrefaction itself
was an instance of spontaneous generation of life too small to be seen.

   The argument began when syphilis was first introduced into Europe after
Columbus' voyages to the New World. Girolamo Fracastoro (1478–1553),
who practiced medicine in Padua and Verona, was a classmate of Copernicus.
He took an early interest in the causes of disease. He named syphilis and wrote
up a detailed medical description of it in a 1530 poem "Syphilis sive Morbus
Gallicus" in which the afflicted character in the poem, a shepherd, was named
Syphilus. The poem described Fracastoro's opinions on the origins, symptoms,
and treatment (mercury) of the disease. In 1546, he published a scientific treatise
on the disease *De contagionibus et contagiosis morbis et eorum curatione*. He also
developed an interesting idea. He said that contagious diseases were caused by
tiny spores or seeds, seminaria, too small to be seen. These seeds moved in the
air and entered the body by way of the breath or the bloodstream, leading to
infecting disease. His spore hypothesis, published in 1546, was just that.[5] There

Francesco Redi was an Italian scientist who received his medical education in Pisa. In 1668, he published *Experiments in Insects* for which he is considered the founder of experimental biology. He showed that maggots did not arise from rotting meat but from eggs laid on the meat. He used jars with meat, either covered with gauze or uncovered, and thus offered the first controlled experiments to prove what should be called the Redi Doctrine: All life arises from preexisting life. He also was a founder of the field of parasitology, identifying 180 species of parasites. He classified worms into three groups—annelids (earthworms), helminths (tapeworms), and nematodes (roundworms like pinworms or *Ascaris*), although those names were not introduced for another century.

were no microscopes in the early 16th century. Most of his contemporaries believed contagious diseases were caused by foul odors or miasmas. They, too, had no evidence to identify such miasmas as causes. What they had was correlation. Illnesses were common in the less-cared-for neighborhoods among the poor. That included less personal hygiene than the wealthier could afford and more sewage in the streets with its accompanying foul odors.

It became a debate, two centuries later, when John Turberville Needham (1713–1781) and Lazzaro Spallanzani (1729–1799) performed experiments on putrefaction.[6] In Needham's demonstration, crushed wheat was placed in water and boiled. Crushed wheat that was not boiled, he noted, very quickly spoiled, with the appearance of a foamy surface and cloudy color of the water. But crushed wheat that was boiled and then allowed to stand also spoiled; but it took longer for this to appear. Needham claimed this was attributable to a "life force" that entered from the air. Boiling dissipated that life force, but if exposed to air the life force would return. If the jar with the wheat and boiled water were sealed after boiling, the spoilage usually did not take place. Needham believed that was because the life force that was spent could not be replaced from the air.

Spallanzani used meat broth and did a similar set of experiments. He was educated by Jesuits and then studied law. Then his cousin Laura Bassi at the University of Bologna introduced him to science. She was a proponent of Newtonian physics, the second woman ever to earn a doctoral degree, and the first woman to earn a professorship at a European university. Broth that was exposed to air, boiled or unboiled, spoiled. The broth that was sealed in a jar but not boiled also spoiled. The broth that was boiled and then sealed did not spoil. To Spallanzani, this was evidence that Fracastoro's spore hypothesis was correct and that a living agent, not some vague vital force, was present in the air. To Needham, it was still the vital force and Spallanzani had not shown the spore hypothesis. Both Needham and Spallanzani were Roman Catholic priests. Neither scientist looked on the nature of putrefaction as a test of God's role in creating or maintaining life. During this 18th century, however, there was a growing belief among deists and atheists that spontaneous generation was compatible with either a godless universe or a god who operated by the laws of the universe and not by miracles or the supernatural. Those who held the belief that life had an uninterrupted history that dated back to the creation described in Genesis looked on spontaneous generation as inherently false.

Louis Pasteur Shows Microbes Arise from Preexisting Microbes

A settling of the Spallanzani–Needham dispute was performed by Louis Pasteur. He started out as a chemist, and for his dissertation he showed chirality (handedness) in molecules by using tweezers to separate organic crystals (tartrates) and demonstrating microscopically that they had a left- or right-handed orientation. He used his microscope to study the causes of diseases in silkworms and the spoilage of beer and wine, which sometimes turned to vinegar. He concluded these were caused by microorganisms, and once again his microscope showed healthy yeast in beer and wine formation were replaced by much smaller rod-shaped bacteria in the spoiled vats and barrels of the beer makers and winemakers. In 1862, Pasteur devised a way to distinguish Spallanzani's and Needham's experiments. He prepared a flask with a spigot curved like a swan's neck and he boiled a meat broth. The U-shaped curvature of the neck did not allow microbes to pass through to the cooling broth. It remained clear for days. He then showed if he tilted the flask to let some of the broth enter into the U-shaped spout and then return to the flask of broth, the broth would spoil. He also showed that under the microscope there were no bacteria present in the sterile broth, but in the spoiled broth it was teeming with bacteria.[7]

Pasteur's work established the biogenetic law that Redi had first proposed: All life arises from preexisting life. Pasteur also showed, with a series of flasks containing sterile broth, that such bacteria were present in the air on mountains and at sea far from land. Fracastoro's spores were no longer some sort of chemical or vague agent that caused spoilage; they were actual microorganisms consistent with the findings of Antonie von Leeuwenhoek, who had first shown such life teeming in the fluids he sampled with his microscopes.

Wilhelm Hofmeister Introduces Alternation of Generations in Cryptogamic Plants

The life cycle implies a repetition. That makes it profoundly different from spontaneous generation in which there is no hint from the material involved of what organisms will arise from old clothes, stored paper, aging fruits and vegetables, or aging meats. There is also no clue to what process goes on in such wastes and spoiled foods that generate complete adult organisms. But a life cycle is filled with specific stages that can be accurately described. The early stages, of course,

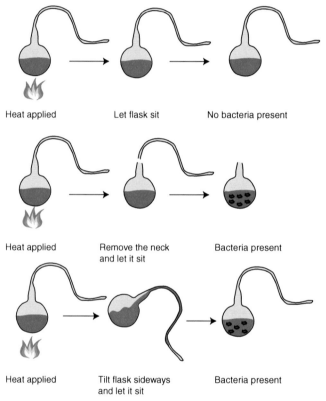

Heat applied	Let flask sit	No bacteria present
Heat applied	Remove the neck and let it sit	Bacteria present
Heat applied	Tilt flask sideways and let it sit	Bacteria present

Louis Pasteur used flasks with swan- or goose-neck-shaped tubes. He took three such flasks containing broth and heated them to kill all life in them. The first he allowed to cool without disturbing the flask. It did not convert the broth into a murky condition filled with bacteria. The second he broke off the neck near the flask and it quickly became corrupted with bacteria. In the last flask, he allowed the sterile broth to cool and then tipped the neck so fluid fell into its U-shaped area and then allowed the broth to run back into the flask. It too became clouded with bacteria like the first flask. Pasteur claimed the bacteria had fallen into the outer bend of the U-shaped tube and only when tilted and run back to the flask could the bacterial contamination occur. This experimental evidence convinced most scientists that he had confirmed Redi's Doctrine: All life arises from preexisting life.

could not be described until the advent of microscopy. As scientists used these tools for preparing specimens and examining them under microscopes they made many new findings. Wilhelm Hofmeister (1824–1877) made such observations and published these in 1851.[8] He was the son of a publisher in Leipzig and used his own press to publish his work. He took particular interest in the ferns, mosses, and bryophytes. These were less studied than the flowering plants. Botanists collectively referred to these nonflowering forms as part of "cryptogamic" botany. Their reproductive organs were largely unknown or unseen unlike the showy sexual display of flowers. Hofmeister was surprised by his findings. The cryptogamic forms (mostly mosses and bryophytes) shared an unusual life cycle in which they lived out two different plant lives. One, usually smaller, form of the plant was capable of producing reproductive cells. These united to form the larger plant, which bore a fruit body filled with spores. The gamete-producing, or sexual, stage was called the gametophyte and the spore-bearing, or nonsexual, stage was called the sporophyte. Hofmeister described this as an alternation of generations. In a convoluted way, we can say that we have an alternation of generations, but our gametic stage would be limited to the events taking place during egg formation in ovaries or in the Sertoli cells of our testes where mature spermatozoa form. Our sporophyte equivalent in humans begins with fertilization of the egg by a sperm.

Through Hofmeister's careful studies of cryptogamic plant species, the duality of our life cycle emerged; however, Hofmeister did not know that the alternation of generations was related to the chromosomal composition of those cells. Mitosis and meiosis would be worked out after Hofmeister died in 1877.[9] The chromosome number in the sporophyte was twice that of the chromosome number in the gametophyte. The gametophyte plants were called haploid. The sporophytes were called diploid. The gametophyte sex cells were haploid but produced by mitotic division in specialized parts of the gametophyte. The sporophyte spores were haploid, produced by meiosis in the sporangium of that larger plant. They formed the gametophyte male or female haploid plants. The male-bearing plant produced motile sperm-like cells. The female-bearing plant housed an egg in a receptacle awaiting the entry of a motile sperm-like cell. This might be represented as diploid sporophyte produces haploid spores by meiosis → spores and resulting gametophyte are haploid → gametophyte produces male or female gametes by mitosis → fusion of male and female gametes creates diploid zygote → zygote develops into new sporophyte.

The Constancy of Chromosome Number Becomes the Theoretical Basis for the Life Cycle

We can now add a layer of information to the life cycle that would have been unimagined before the advent of microscopy. The genetic significance of the life cycle is the repetition of the constancy of the chromosome number for the species each generation. Some species like yeasts can produce a life cycle in which half the cells are haploid and half the cells are diploid, with the occasional union of two haploids to produce a new generation of diploids. The haploid or diploid predominance may be associated with the environmental conditions. In the cryptogamic plant species, some have both stages equal in duration. Some have a short gametophyte stage and a long sporophyte stage. Some have the reverse with a short sporophyte stage that serves to produce a flood of gametophyte plants. In humans, the haploid stage (i.e., the gametes) is not only hidden

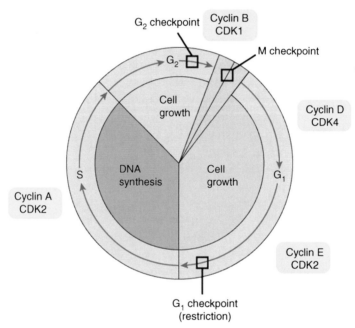

The cell cycle when I learned it as a graduate student about 1957 was $G_1 \rightarrow S \rightarrow G_2 \rightarrow M$ and then back to G_1 again. Before 1950, biologists believed there were stages of mitosis—prophase, metaphase, anaphase, and telophase—followed by a long interphase and then another round of mitosis. The G stands for "gap" because before the 1970s, the genes involved in the cell cycle were unknown. As they were worked out, it led to a richer understanding of the role of the cell cycle and metabolism. In most human tissues (muscle, bone, tendon, fat, nerves), cells remain in the G_1 stage for most of our adult lives. For some tissues (skin, blood, the lining of the intestines, reproductive tissue), cells divide for most of our lives, repeating the cell cycle each time. For differentiated, adult, nondividing cells, some biologists use the G_0 state as a terminal state in which the cell has lost the capacity to shift into cell division. Some mutations in cell cycle genes lead to cancers. Other cell cycle mutations can lead to premature aging or self-destruction (a process called apoptosis, or programmed cell death).

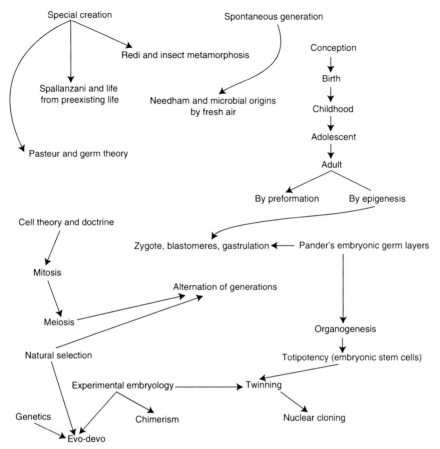

Until the Renaissance, most of humanity thought of visible life as arising from a special creation by supernatural means. Others believed that lower forms of life arose from spontaneous generation, especially from rotting, decaying, or spoiled garbage. For human origins, they recognized the need for insemination by a female (called conception) leading to birth. The resulting child became an adolescent and then an adult. Redi and Spallanzani provided experimental evidence for the continuity of life from preexisting life. Diehards for spontaneous generation used Needham's evidence for fresh air to activate dead material and restore life. Pasteur's experiments proved convincing that even microbial life required continuity from preexisting bacterial or yeast cells. A similar debate occurred over how organisms shifted from conception to birth. Preformationists believed growth was the mechanism, and some proposed that all future descendants were present in the eggs or sperm of a living person. Opposed to this were those who followed Aristotle's view that adult organs arise by epigenesis, an emergence requiring an unknown modification of simpler stages into complex organs. The cell theory and doctrine led to the discovery of mitosis, meiosis, and gamete formation, resulting in an understanding of fertilization and zygote formation. It led to the epigenetic changes in embryonic layers, resulting in processes such as a neural plate leading to a neural tube leading to cephalic (brain) and posterior (spinal cord) formation. Experimental embryology led to an understanding of identical twins from portioning of blastomeres in early embryos. It led to an understanding of chimerism from fusion of zygotes. It incorporated developmental mutations (operons, homeotic mutations, homeoboxes). It also connected comparative anatomy and evolution at a molecular level in the new field of evo-devo.

at a cellular level from our perception, it is limited to that part of our life cycle from puberty (and menarche) to menopause in women and from puberty to death in males, although the sperm count dwindles with old age and senescence.

The idea of life cycles suggested to microscopists in the mid-20th century that there was a corresponding cell cycle. For the mitotic process, this was divided into four stages: an M or mitotic stage in which the chromosomes coiled, become compacted, are pushed into an equatorial plane, separate at their centromeres, get distributed to poles of the cell, and are re-encased in nuclear envelopes as the cell cytoplasm partitions into two. This is followed by a G_1 stage in which the cell carries out its metabolic roles and enlarges. The chromosomes are unwound and genes are turned on or off depending on the tissue function and stage of the life cycle. The third stage or S stage, involves the synthesis of DNA in the chromosomes of the nucleus. This is followed by the G_2 stage in which the cell metabolism shifts to production of the components needed for a successful mitosis or M stage. The genetic control of these cell cycle events was initiated by Potu N. Rao (b. 1930) and R.T. Johnson in 1970 using cell fusion techniques with cells in different stages of the cell cycle. This was later analyzed by studies of yeast cells showing specific genes involved in these stages of the cell cycle and the proteins that regulated, stopped, or initiated these events. The work of Paul Nurse (b. 1949), Leland Hartwell (b. 1939), and R. Timothy Hunt (b. 1943) identified the proteins and genes that control the cell cycle.[10]

The Cell Cycle Is the Predecessor to the Life Cycle

Note that two major cycles are involved in the multicellular organism. There is the life cycle, but that would not be possible without the cell cycle to multiply the cells. And in most multicellular species, there is a third cycle involving the production of gametes through the process of meiosis, a much-modified cell cycle that requires two cell divisions for completion as diploid reproductive cells (in humans $2N = 46$) are converted into mature haploid cells (in humans $N = 23$ as sperm or eggs). The life cycle is a broad one encompassing the entirety of each species' way to perpetuate itself. This idea required the elimination of the widely held belief in the spontaneous generation of lower life-forms. The mitotic cycle is limited to the multiplication of cells in the life cycle. The meiotic cycle is limited only to that phase of the life cycle in which reproduction by formation and union of gametes takes place. The repetition of the stages of the life of humans experienced by reflective poets or philosophers through the ages was largely void of knowledge of the underlying biology. The biology added to and infused the poetic images of our paths through life, as experimentation, theory, and new technologies permitted this analysis and refinement to take place.

References and Notes

1. Dublin L, Lotka A. 1936. *Length of life: A study of the life table.* Ronald Press, New York. Also, Oeppen J, Vaupel JW. 2002. Broken limits to life expectancy. *Science* **296:** 1029–1031.

2. Steinbock B. 1992. *Life before birth: The moral and legal status of embryos and fetuses.* Oxford University Press, New York.

3. McCartney ES. 1920. Spontaneous generation and kindred notions in antiquity. *Trans Amer Philos Assoc* **51:** 101–115.

4. Redi F. 1668. *Esperienze intorno alla generazione degli insetti.* [*Experiments on the generation of insects.*] Trans. Bigelow M. 1909. Open Court Publishing, Chicago. Also see wembryo.asu .edu/pages/francesco-redi-1626-1698/; www.scientus.org/Redi-Galileo.html/.

5. Fracastoro G. 1546. *De contagione et contagiosis morbis et eorum curatione. Libri III.* [*On contagious diseases and their cure.* 3 books.] Trans. Wright WC. 1930. Putnam's Sons, New York.

6. Spallanzani L. 1765. *Saggio di osservazioni microscopiche concernenti il sistema della generazione dei Signori di Needham e Buffon.* Eredi di Bart. Soliani, Modena. Also see Capanna E. 1999. Lazarro Spallanzani: At the roots of modern biology. *J Exp Zoo* **285:** 178–196.

7. Pasteur L. 1922. *Oeuvres de Pasteur. Tome II: Fermentation et Générations dites Spontanées (1860–1866)* (ed. Vallery-Radot P), pp. 185–358. Masson et Cie, Paris. Also see Roll-Hansen N. 1979. Experimental method and spontaneous generation: The controversy between Pasteur and Pouchet, 1859–64. *J Hist Med Allied Sci* **34:** 275–292.

8. Kaplan DR, Cooke TJ. 1996. The genius of Wilhelm Hofmeister: The origin of causal analytic research in plant development. *Amer J Bot* **83:** 1647–1660.

9. Voeller B, ed. 1968. *The chromosome theory of heredity: Classic papers in development and heredity.* Appleton-Century-Crofts, New York.

10. Nurse P. 2000. A long twentieth century of the cell cycle and beyond. *Cell* **100:** 71–78. Also, Rao PN, Johnson RT. 1970. Mammalian cell fusion: Studies on the regulation of DNA synthesis and mitosis. *Nature* **225:** 159–164.

The Molecular Basis of Life: From Vitalism to Organic Molecules to Macromolecules

Protoplasm as the basis of life. Darwin's warm little pond, coacervates, Oparin's soup. Proteins as enzymes, Garrod's inborn errors of metabolism, biochemical pathways, molecular disease. DNA as genetic material via transformation, DNA as genetic material via phage infection, DNA as a double helix, replication of DNA, protein synthesis. Regulatory genes as operons, homeobox genes and development, molecular phylogeny, synthetic genes, synthetic life.

It is astonishing how fast the past is forgotten or displaced. When I was an undergraduate at New York University, students took a course in zoology and a course in botany. There was no biology course except in high school. When I was a graduate student studying genetics with Nobel geneticist H.J. Muller, I was in a zoology department. When I was an instructor at Queen's University in Kingston, Ontario, there was a biology department, but biochemistry was not taught in the biology department. It was taught in the medical school because biochemistry was considered part of medicine. I returned to a zoology department as a young professor at the University of California, Los Angeles. When I left after 8 years to join the faculty at Stony Brook University, the biology department had just split into a department of biochemistry and a department of biology. A year later, the biology department split again and a separate department of ecology and evolution was formed. The remnant of the biology department remaining was renamed a department of comparative biology. Finally, the department of comparative biology went extinct and each of us in it was told to seek a new home among the two remaining departments or among departments in the medical school. I felt like baseball players before the 1960s when a player never knew when he would be traded and had no power to prevent it and no authority to leave and seek employment in another major league team.

The Growth of New Fields of the Life Sciences

New fields emerge by fusion and splitting as they grow in influence or fade into the past. Old fields reemerge, sometimes with new names. The changes reflect the prevailing views of how the life sciences should be partitioned. Sometimes those changes are initiated by a department or a significant core of its members. Sometimes the change is imposed by a dean or a provost. What influences the reassignment of members or the names of departments is the way scientists depend on each other for collaboration, criticism, or the potential to attract students. The new attracts more effectively than the old. It also makes a difference in the likelihood of obtaining federal grants to support research. The idea of such support is relatively recent because major granting agencies in the United States, like the National Science Foundation or the National Institutes of Health, are creations that emerged after World War II or after Sputnik, the USSR space vehicle that shattered American esteem in 1957 and led to a flood of federal funding to generate more scientists. Since the 1890s, the hallmark of biological research has been experimental approaches to teasing out nature's mechanisms and testing theories about how life works. Whereas wealthy scientists like Darwin or Galton could support their own research, most scientists could not. We have forgotten that Morgan and his students were supported in part by Morgan's generosity in buying the bananas and his students' understanding that they were to bring in their cream bottles to the laboratory for use in the fly work; if they brought in several bottles each day, as Sturtevant related, no question was raised about where they came from. Departmental budgets for research existed, but they were modest.

Biochemistry Is Introduced to Biology in the 19th Century Replacing Vitalism

Biochemistry is a relatively young field of biology. It was virtually nonexistent before the 1800s. Chemistry as a science had barely emerged from alchemy in the early 1700s. The atomic theory of chemical combination to produce molecules was introduced in 1805 by John Dalton (1766–1844). Dalton had distinguished elements (consisting of a uniform number of atoms of the same weight) and molecules (compounds of two or more atoms of different elements).[1] Until 1828, it was universally believed that living matter involved some vital force that was endowed by God at the creation of life in Genesis. The assumption relied on the belief that God would not make known to humans how components of life worked or how they were composed for fear that humans would use that knowledge to synthesize life and act like gods. The "tree of life" that was forbidden to Adam and Eve in the Garden of Eden was used to justify this belief. Those sharing this view are called vitalists, whether their motivation arises from religious tenets or not. They believe in a uniqueness of living material that defies chemical or physical analysis (reductionism) to components that can then be reassembled and function.

The first debate over vitalism in biology concerned static electricity. Luigi Aloysii Galvani (1737–1798) was studying the muscle and skin of frogs. He and an assistant had produced static electricity in the form of sparks by rubbing dried frog skin with a rod. When the rod touched the dead frog's leg, it contracted, much to the surprise of Galvani and his assistant. Galvani assumed that the spark came from the once living matter (the skin that was rubbed) and that it activated the electricity that was in the dead frog's muscle. He called this animal electricity. When he reported his work to his colleague Alessandro Volta (1745–1827), Volta disagreed. He believed from his own work that electric sparking was a physical phenomenon not uniquely tied to life.[2] To prove his point, Volta constructed the first inorganic battery in 1799. He stacked zinc, copper, and electrolyte fluid and discovered that electricity could be generated chemically (known then in his honor as a Voltaic pile). The idea of electricity producing lifelike behavior in dead frogs swept across the popular imagination in Europe, and it led to studies of electricity and its effects on living organisms, including humans willing to participate in such experiments. It also led, in the popular literature, to assigning electricity as the basis of animating Dr. Frankenstein's monster assembled from several corpses. Mary Shelley's *Frankenstein, or the Modern Prometheus*, was published 1818. In 1816, Shelley was known to have discussed Galvani's experiments as well as the possibilities of reanimation with Percy Shelley and Lord Byron, although her novel does not specify Dr. Frankenstein's methods.

Freidrich Wöhler Synthesizes Urea from Ammonium Cyanate

The attack on vitalism continued once electricity was abandoned as the source of animation in living tissue. Vitalists retreated from living behavior to the composition of life and held firmly to denying a chemistry of life was possible. That was challenged by Friedrich Wöhler (1800–1882), who studied medicine in Heidelberg and then switched to chemistry and studied with Jöns Jacob Berzelius (1779–1848) in Sweden. In 1824, Wöhler showed that oxalic acid could be synthesized from cyanogen, an inorganic compound. Oxalic acid is not as well known a compound as the one Wöhler synthesized 4 years later. His 1828 paper showed how urea could be synthesized from ammonium cyanate, an inorganic compound.[3] Wöhler also studied meteorites and showed that some of them contained organic compounds, raising the possibility that life might exist elsewhere in space. Urea, $CO(NH_2)_2$, was first isolated in 1773 by Hilaire Marin Rouelle (1718–1779) from urine, although some sources say Herman Boerhaave (1668–1738) was the first, isolating urea in 1727. Wöhler compared his powdered urea, prepared from ammonium chloride and potassium cyanate, with the urea isolated from urine and showed they were identical. Wöhler was trying to synthesize ammonium cyanate (he began these syntheses in 1823 and ended them in 1828) and wound up with urea. Wöhler wrote to Berzelius that "...I can make urea without needing a kidney, whether man or dog. The ammonium salt of cyanic acid is urea." His quote emphasizes the point that an organic compound was synthesized outside a living system, thus defeating the idea of vitalism. Skeptics in favor of vitalism remained, including Justus von Liebig (1803–1873) and Louis Pasteur, both carrying considerable esteem as chemists. But by 1855, many organic compounds had been synthesized by chemists from inorganic compounds and the field of organic chemistry was launched. Vitalism shifted once again and moved into protoplasm, the sarcode or living matter identified in the cell in mid-19th century.

Fermentation Becomes a Biochemical Process

The reductionist's study of life shifted also and moved into evolution in 1858 and into the cell in mid-19th century (1830–1880). It shifted to the nucleus of the cell from the 1870s to the 1890s. The suspicion of the nucleus as playing a significant role in cell division, the preservation of the chromosome number, the formation of gametes, and the possibility that they contained hereditary factors was finding publication as speculative articles in journals and in books. These publications also stimulated the interest of new generations of chemists. Reductionism also shifted to protoplasm and its capacity to carry out metabolic reactions. When Louis Pasteur proposed his germ theory of infectious diseases, he based it on his prior work that showed that yeasts were involved with fermentation of sugars and other carbohydrates to produce alcohol. He also showed that spoilage of food (putrefaction) was associated with bacterial action, and beer and wine growers' contaminated barrels and vats harbored smaller rod-shaped bacteria that turned the grape juice or cereal mash into vinegar.[4]

Pasteur assumed that the yeast cells had substances within them called "ferments" that brought about this process. He was a vitalist and had destroyed the view that spontaneous generation produced any living forms. Living cells had unique living properties that included the capacity for fermentation, spoilage, and disease. A somewhat different approach was developing in the laboratory of Ernst Felix Hoppe-Seyler. He obtained his M.D. in Berlin and then became an assistant for Rudolph Virchow at the Berlin Pathological Institute. Virchow's studies of cells and cellular pathology appealed to him, and he shifted to study the properties of cells from the chemist's approach. Beginning in 1857, he worked out the relation of hemoglobin to oxygen and showed that oxygen bound to that molecule. He crystallized the purified hemoglobin he obtained. He also showed that hemoglobin molecules carried iron. His work led to the formation, in Strasbourg, of the first biochemistry department (called physiological chemistry then), and he was made the head of that department.[5] Hoppe-Seyler later inspired research on nucleic acids that led to the isolation and chemical characterization of DNA by his students Friedrich Miescher and Albrecht Kossel.

Buchner's "Press Juice" Uses Reductionism to Demonstrate Biochemical Fermentation

Eduard Buchner (1860–1917) took an interest in Pasteur's findings of fermentation, but he favored Hoppe-Seyler's approach to that of the microscope. He took a healthy suspension of yeast cells and ground them up in a mortar and pestle with fine sterile minerals crushed to a powder. He then poured this mixture into a press and squeezed out the clear liquid. He called this "press juice" and applied it to a water solution with sugar. Very soon bubbles of carbon dioxide began to fill the jar and alcohol was produced, probably by glycolysis. Buchner proved that chemically isolated components of cells and not the cells themselves were involved with the fermentation process. The ferments he obtained were the actual cause of the metabolic change.[6] Buchner erred, however, in assuming that in the living condition the yeast cells released these ferments into the surrounding liquid to bring about fermentation. Later work showed that the process was the reverse; the sugar solution entered the yeast cells where the fermentation, or glycolysis, took place. Wöhler's work showed that organic compounds found in living things could be synthesized without the presence of living cells. Buchner's work showed that processes associated with living cells, such as fermentation, could also be performed by chemists without using live cells to do the same thing. Such cell-free fermentation was another blow to vitalism, which kept being pushed out of each new location or process found in living organisms.

Miescher's Pus Cells and the Origin of Nucleic Acids

One of Hoppe-Seyler's students was Johannes Friedrich Miescher, a Swiss student born into a family of scientists. He studied medicine but had poor hearing, so he decided on a research career instead. He thought he might investigate phosphorous compounds and he sought them in pus cells. These were mostly white blood cells (leukocytes), and he could obtain large numbers from the bandages removed from infected wounds. He extracted a purified preparation of nuclei and then precipitated out of the ruptured nuclei an acidic material that he called nuclein.[7] Miescher recognized that the presence of nuclein in the nucleus was important and noted a chemical difference between the nucleus and cytoplasm, but did not understand the ultimate significance. Later, another of Hoppe-Seyler's students, Albrecht Kossel, prepared nuclein extracts and isolated what he called nitrogenous bases, later called purines (adenine and guanine) and pyrimidines (thymine, cytosine, and uracil). Nuclein then became nucleic acid.[8] This was still biochemistry. The biochemical basis of life was greatly extended throughout the 20th century with identifications of ferments as enzymes and enzymes as proteins. Vitamins and hormones were isolated and their functions identified. Lipids and carbohydrates could be distinguished chemically from proteins and nucleic acids. Nucleic acid from cells was shown to be composed of two forms, deoxyribonucleic acid and ribonucleic acid. Deoxyribonucleic was primarily found in nuclei, and ribonucleic acid was more likely to be present in the cytoplasm. Metabolic pathways were worked out first by biochemists and then by biochemists working with geneticists.

Molecular Biology Gets Its Name in 1938

What is called molecular biology was first named in 1938 by Warren Weaver (1894–1978). It relied on several cognate fields.[9] Biochemistry was certainly one tributary feeding into this new field. A second was theoretical inference based on the biology of cellular processes. A third was the introduction of X-ray diffraction techniques to study crystalline structure of molecules. A fourth was the development of tools to show the existence of macromolecules, which were smaller than cell organelles but larger than the biochemist's working molecules. Molecular biology was created in a two-pronged effort to explore the structure of cell macromolecules—proteins and nucleic acids. Proteins were the first target. Biochemists showed that proteins were composed of amino acids. There were about 20 of these found in all living cells from bacteria to humans. These were arranged in strands. Smaller strands of amino acids were called peptides. Some of these peptides were bundled and connected by special amino acids bearing sulfur, and these sulfur bonds could only be broken by high temperatures (such as boiling) or chemical reaction. The larger proteins folded into complex shapes but were essentially one long strand. Linus Pauling (1901–1994) played a major role in interpreting the structure of proteins with X-ray diffraction and his discovery that hydrogen bonding formed to produce folded sheets and helical symmetry for many of these segments of the protein molecule.[10]

Molecular Structure Is Revealed through X-Ray Diffraction Studies

X-ray diffraction was introduced into chemistry by William Henry Bragg (1862–1942) and his son William Lawrence Bragg (1890–1971). They applied the work of Max von Laue (1879–1960), who in 1912 showed that X rays could be diffracted by crystals. The Braggs found that X rays directed at a crystal

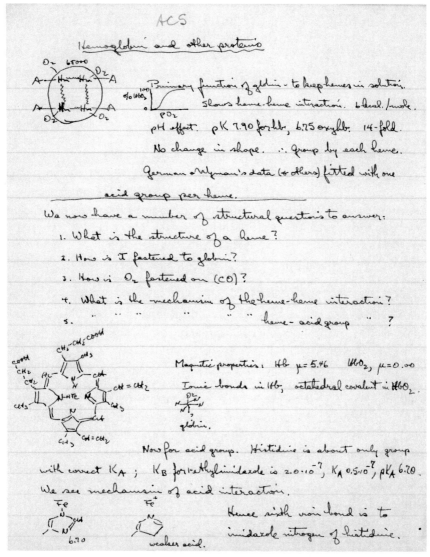

Linus Pauling's notes on the structure of the hemoglobin molecule show how he conceived the structure of the molecule. In the *upper left*, he shows how four heme groups could be organized by the surrounding protein components of the molecule. In the *middle*, he lists a set of tasks to solve the relation of the protein components to the heme and how the molecule carries and releases oxygen. The complex structure in the *lower left* is his visualization of how the heme molecule could be organized. Later, Pauling used normal and sickle cell anemia hemoglobin to work out the defect as a single amino replacement in one of the protein subunits of hemoglobin. He characterized this analysis of sickle cell hemoglobin as evidence it was a "molecular disease." Note the Baconian way Pauling uses data to generate ideas or theories rather than shoring up a theory with data.

produced a diffraction pattern that could be recorded on film; this diffraction pattern showed how the atoms in the crystal were aligned because the path of the X ray was bent according to the way the individual atoms formed their associations in the crystals.[11] They used table salt (NaCl), zinc sulfide (ZnS), and carbon atoms aligned in diamond. Bragg's student John Desmond Bernal (1901–1971) used X-ray diffraction to study steroids and their parent molecule, cholesterol, and worked out their structures. He also used X-ray diffraction to work out the structure of vitamins, and his most ambitious project was to study the structure of a virus, tobacco mosaic virus.

Another of Bragg's students was William Astbury (1898–1961), who got a position later at Leeds where he studied the proteins associated with wool (which has a lot of keratin) and collagen. He found these molecules when wet rather than dry showed a considerable capacity for stretching, and he identified that with the helical nature of their molecules. His X-ray diffraction confirmed this helical organization. Astbury also received a sample of DNA crystal in 1937 from Torbjörn Caspersson (1910–1997) in Sweden. Astbury's studies showed it had a repeating unit and the nucleotide bases were oriented in a common flat direction, but he could not work out a satisfactory structure for the model.[12] Astbury published his first X-ray diffraction photo of DNA in 1938. James Watson was at a 1951 meeting in Naples that included Maurice Wilkins (1916–2004) whose DNA photo excited Watson's interest. In *The Double Helix* Watson refers to an earlier "one half good photo in the published literature" by Astbury, probably the 1938 photo. Watson hoped he could work with Wilkins but, as fate often happens, he found his mentor for X-ray studies of DNA with Francis Crick. Astbury's work was very influential on Linus Pauling, who confirmed the helical organization of collagen and extended it to other protein molecules. The α-helix was a more precise characterization, and Pauling showed how the hydrogen bonds suggested by Astbury in 1931 were related to both the α-helix and the folded sheets that Pauling found for proteins a generation later.

It is typically the fate of most scientists to have their works forgotten or eclipsed by the more detailed and informative work of others. Many people have their names associated with the working out of the structure of DNA, but Astbury is almost never discussed. His work appeared about 15 years before the double helix model was worked out.

Muller and Schrödinger Provide a Theory of the Gene as a Crystal

The theoretical tributary of molecular biology came from the work of H.J. Muller and Erwin Schrödinger (1887–1961).[13] Muller made the study of mutation and

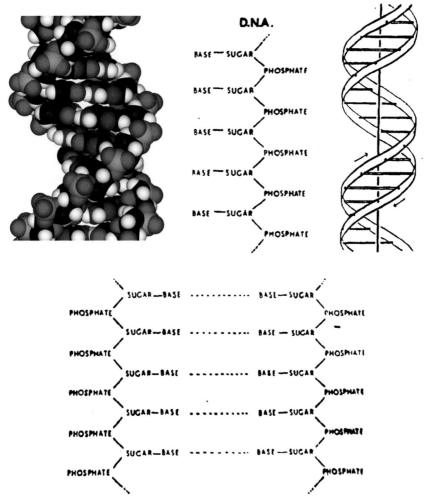

Four ways to visualize the double helix model of DNA. The *top left* image shows a model with the atoms in place. The *top middle* image shows the sugar-phosphate bond arranged in a single strand of DNA. The *top right* image shows the double helix as two helical strands moving in opposite directions. The *bottom* image is a two-dimensional representation showing the base pairing as rungs within the sugar-phosphate outer "ladder." The area for hydrogen bonding is represented by the horizontal dotted lines. In April 1958, I was asked to host Linus Pauling while he spent five days giving lectures on the UCLA campus. I had just bought a copy of *The Double Helix* book that came out that day. Pauling asked to flip through it and I asked him if he would have been able to infer the structure of DNA if he had seen the photo that Watson and Crick used taken by Rosalind Franklin and Ray Gosling. He said, "Science does not work that way. Sometimes you have a mindset that doesn't see the obvious." I always respected Pauling for his integrity after hearing that response.

the gene his life's work. He had characterized the gene as being the basis of life because it had the capacity of replicating its mutations. In 1921, when speaking at an American Association for the Advancement of Science (AAAS) meeting on "Variation due to change in the individual gene," Muller suggested that the newly discovered bacteriophages (then called d'Hérelle bodies) were similar to genes in being noncellular and that their properties were hard to distinguish from that of the gene. He concluded with an inspired insight that "...if these d'Hérelle bodies were really genes, fundamentally like our chromosome genes, they would give us an utterly new way to tackle the gene problem. They are filterable, to some extent isolable, can be handled in test tubes, and their properties, as shown by their effects on the bacteria, can thus be studied after treatment. It would be very rash to call these bodies genes, and yet at present we must confess that there is no distinction known between genes and them. Hence we cannot categorically deny that perhaps we may be able to grind genes in a mortar, and cook them in a beaker after all. Must we geneticists become bacteriologists, physiological chemists, and physicists, simultaneously being zoologists and botanists? Let us hope so."[14]

Muller extended his theoretical insights in 1936 after a short stay in Berlin where he worked with Nikolay V. Timoféef-Ressovsky (1900–1981) and after meeting biophysicist Karl Günther Zimmer (1911–1988) and physicist Max Delbrück, who expressed interest in studying the gene from the perspective of physics. After Muller left for the USSR, they attempted to measure gene number

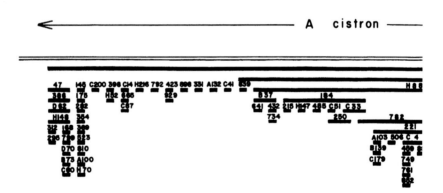

Seymour Benzer (1921–2007) used bacteriophage T4 to study a region called the rII gene. Benzer worked out a saturated map that revealed the region consisted of tandem genes, rIIA and rIIB. He called them cistrons, a term he coined to represent a gene by its individual function. This map reveals that most of the mutations he studied were single point mutations, and a few were larger-sized deletions of two or more sites in the recombination map. Genetic fine structure suggested to most geneticists that Benzer's recombination study of induced and spontaneous mutations in the rII region were consistent with a DNA sequence of several hundred base pairs.

and size by target theory and wrote up their work in 1935. Muller's 1936 paper bore the title "Physics in the attack on the gene."[15] He mentions the work of Astbury and the significance of using biologically significant molecules for this analysis and hoped that it could be extended to the newly found salivary chromosomes that were bundles of hundreds of duplicated threads. He compared the gene with a type of crystal that had the unusual property of retaining its capacity of replication while being altered by new mutations.

Muller's views were picked up by Erwin Schrödinger through the Timoféef-Ressovsky, Zimmer, and Delbrück paper, which was known as "The Green Pamphlet" or "The Three-Man Paper."[16] In it, the investigators wrote, "We view the gene as an assemblage of atoms within which a mutation can proceed as a rearrangement of atoms or a dissociation of bonds... ." Schrödinger included a discussion of this paper in *What Is Life?* Muller was not directly cited for his 1921 beliefs that the gene might be subject to a physicist's analysis. Similarly, Muller's 1936 paper was not likely to have been read by Schrödinger. Schrödinger adopted the ideas of the gene as a crystal, but he added several brilliant insights.[17] He first designated the gene as an "aperiodic crystal" that retained its capacity to replicate but allowed for change in its structure that made it aperiodic. He also believed that the gene as an aperiodic crystal might have a codescript and this information or specificity would be translated in the cell into its varied activities. He claimed the codescript could be quite simple and offered the example of Morse code, which uses a combination of dot, dash,

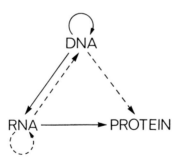

About the time of their 1953 publication of the DNA double-helix model, Watson and Crick were exploring the idea of information flowing from DNA→RNA→Protein (which my fellow graduate students stated as "DNA makes RNA makes Protein"). Crick called this the "genetic dogma" or the "Central Dogma" because experimental evidence for it was lacking. By 1970, work on microbial synthesis, especially viruses, suggested the route between RNA and DNA flowed both ways. Crick believed it might be possible to go from DNA directly to Protein, but this has not been achieved. Crick doubted that one could work backward from Protein to RNA or from Protein to DNA. Some biologists believe a "Lamarckian" flow from environment to Protein to RNA to DNA (or from environment to RNA to DNA) might be possible. Barbara McClintock did not rule out that possibility.

and space to generate all the letters of the alphabet and hence any communication in any language having an alphabet.

Model Building and the Structure of DNA

One last element entered the formation of molecular biology that most scientists equate with the publishing of the double helix model of DNA in 1953. It is appropriate that the word model is used for the double helix because two of its major properties could not be revealed by X-ray diffraction. One was the aperiodic sequence that was demanded for a gene for its specific function. The second was for the way the nitrogenous bases were associated. Watson

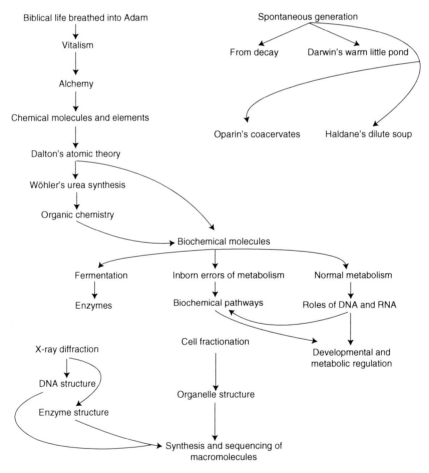

See figure legend on following page.

and Crick produced very accurate cutouts from cardboard of the structures of adenine, guanine, cytosine, and thymine. Watson found that only certain pairs would be of identical length and these were the AT (or TA) pair and the GC (or CG) pair. All other combinations either mismatched with their hydrogen-bonding associations or produced abnormal sizes that would not serve the uniform width demanded by the two helices of DNA.[18]

Incremental Progress from Vitalism to Organic Molecules to Macromolecules

What began as a perception of life that arose spontaneously from the trash and warmth of moist unattended discards shifted to a perception of life arising from life and of life possessing a supernatural God-given vital force. That perception of vitalism yielded to scientists who began to explore the cell having a chemistry that was not profoundly different from the inorganic chemicals that were being produced and analyzed after Dalton's atomic theory changed chemistry. Some chemists began to reproduce organic or other compounds that were believed to be unique to living things. This later extended to processes in the cell, like fermentation, which could be reproduced in cell-free experiments. New techniques shifted the way chemists worked, as physicists joined chemists to develop the field of X-ray diffraction. Each increment involved a

---

The biblical account of Adam's life involved a lifeless body into which a "breath of life" was introduced by God. This led to an outlook in the life sciences that is called vitalism. Life was seen as unique and not subject to reduction to chemical or physical processes. A second view favored in antiquity was spontaneous generation of life from decaying or rotting material, especially for lower forms of life. Alchemy yielded some elements and compounds, but often it was accompanied by vitalistic views for their functions. Dalton's atomic theory shifted chemistry from alchemy. Spontaneous generation was used by Darwin for the origin of life coming out of a "warm little pond." The rapid advance of chemistry showed that organic molecules existed and could be synthesized from nonorganic components as Wöhler showed for urea. This led to organic chemistry. Spontaneous generation was used again for Haldane's concept of early Earth having an ocean of dilute organic soup. A similar idea was behind Oparin's coacervate theory of life formation on mudflats. Biochemical molecules led to an understanding of fermentation and the role of enzymes. It led to a study of human metabolic disorders (Garrod's inborn errors of metabolism). It also led to an understanding of metabolism, both in the breakdown of foods and the synthesis of important cell constituents. The tools of X-ray diffraction led to the structural analysis of complex proteins (hemoglobin) and DNA. Cell fractionation allowed an analysis of organelle function at a biochemical level. The operon model allowed regulation of on/off processes in metabolism and in development. All key molecules—lipids, carbohydrates, vitamins, DNA, RNA, and proteins—have been successfully analyzed and synthesized.

theory, a new tool, a new process, or a series of clever experiments. The life inside the cell shifted from being protoplasm, to being a set of processes that performed numerous activities, to containing a set of molecules that could be synthesized by chemists, and finally to three-dimensional molecules that could be crystallized and their shape and chemical composition worked out by molecular biologists. Muller's prophetic vision of a molecular biology of the gene took less than 35 years to be realized.

References and Notes

1. Dalton J. 1808. *A new system of chemical philosophy*. Bickerstaff, London.

2. Galvani [L]A. 1791. *De Viribus Electricitatis in Motu Musculari Commentarius*. Ex Typographia Instituti Scientiarium, Bononiae, Bologna, Italy.

3. Wöhler F. 1828. Ueber künstliche Bildung des Harnstoffs. *Ann Phys Chem* **88**: 253–256.

4. Pasteur L. 1879. *Physiological theory of fermentation*. Trans. Faulkner F, Robb DC. 1910. Harvard Classics, Volume 38. Collier and Son, New York.

5. Hoppe-Seyler F. 1864. Über de chemischen und optischen Eigenschaften des Blutfarbstoffs. *Virchows Arch* **29**: 233–235. Also see Fruton J. *Molecules and life: Historical essays on the interplay of chemistry and biology*. Wiley-Interscience, New York. Also see Noyer-Weidner M, Schaffner W. 1995. Felix Hoppe-Seyler (1825–1895): A pioneer of biochemistry and molecular biology. *Biol Chem* **376**: 447–448.

6. Buchner E. 1897. Alkoholische Gärung ohne Hefezellen (Vorläufige Mitteilung). *Ber Dtsch Chem Ges* **30**: 117–124. Trans. Friedmann HC online at http://bip.cnrs-mrs.fr/bip10/buchner0.htm.

7. Miescher F. 1871. Ueber der chemische Zusammensetzung der Eiterzellen. *Med-Chem Untersuchungen* **4**: 441–460. Also see Dahm R. 2005. Friedrich Miescher and the discovery of DNA. *Dev Biol* **278**: 274–288.

8. Kossel A. 1881. *Untersuchungen über die nucleine und ihre Spatungs Produkte*. K. Trübner, Strasburg. Also see Dahm R. 2008. Discovering DNA: Friedrich Miescher and the early years of nucleic acid research. *Hum Gen* **172**: 565–581.

9. Weaver W. 1970. Molecular biology: The origin of the term. *Science* **170**: 591–592.

10. Pauling L, Corey RB, Branson BR. 1951. The structure of proteins: Two hydrogen bonded helical configurations of the polypeptide chain. *Proc Natl Acad Sci* **37**: 205–211.

11. Bragg WL. 1913. The diffraction of short electromagnetic waves onto crystal. *Proc Camb Philos Soc* **17**: 43–57. See also Jenkins J. 2008. *William and Lawrence Bragg, father and son: The most extraordinary collaboration in science*. Oxford University Press, New York.

12. Astbury WT, Bell FO. 1938. Some recent developments in the X-ray study of proteins and allied structures. *Cold Spring Harb Lab Symp Quant Biol* **6**: 109–121.

13. Carlson EA. An unacknowledged founding of molecular biology: H.J. Muller's contributions to gene theory, 1910–1936. *J Hist Biol* **4**: 149–150.

14. Muller HJ. 1922. Variation due to change in the individual gene. *Am Nat* **56**: 32–50.

15. Muller HJ. 1936. Physics in the attack on fundamental problems in genetics. *Sci Mon* **44**: 210–214.

16. Timoféef-Ressovsky NV, Delbrück M, Zimmer K. 1935. Über die Natur der Genmutation und der Genstruktur. 1935. *Nach Ges Wiss Göttingen Math-Phys Klasse, Fachgruppe* VI, **1:** 189–245. Also see Wunderlich V. 2011. *In commemoration of the 100th birthday of Karl Gunther Zimmer.* Max Delbrück Center for Molecular Medicine, Berlin.

17. Schrödinger E. 1944. *What is life? The physical aspect of the living cell.* Cambridge University Press, New York.

18. Watson JD. 1968. *The double helix.* Atheneum, New York. Also see Watson JD, Crick FHC. 1953. A structure for deoxyribose nucleic acid. *Nature* **171:** 737–738; Watson JD, Crick FHC. 1953. Genetical implications of the structure of deoxyribonucleic acid. *Nature* **171:** 964–967.

CHAPTER 7

Sex Determination: From Wild Guesses to Reproductive Biology

Speculative era, discovery of sexual anatomy, discovery of eggs. Discovery of sperm, fertilization as a union of one egg and one sperm, chromosomes. Haplo-diploidy, environmental determination, sex reversals, hermaphrodites, sex-determining genes, dosage compensation.

When a broad historical survey is made for a topic in science, it is not surprising to find that most of the current ways of thinking about the subject and the myriad of information at our call come from the mid-19th century to the present. Sex determination fits that well. We take it for granted that a fertilization occurs when a male and female engage in sexual intercourse, and virtually everyone thinks of an egg and sperm coming together to start a new individual on the path to development. A more informed segment of humanity will also know that sex determination in humans is associated with sex chromosomes and that males are XY and females are XX. It is difficult to believe that no one at the time of the Civil War knew this. Fertilization involving the union of a sperm and egg was established in 1876.[1] The sex chromosomes were first recognized in 1905 and named in 1906.[2]

Yet people have always been curious about how nature (or God) generates a male or a female infant. The biblical account has a separate creation of the female (Eve) from Adam but not a separate creation of sexes for all the other forms of life that preceded the creation of Adam. By the time the biblical narrative progresses to Noah and his ark, the impression is that a two-sex model is universal, as pairs of all the creatures of the earth are sent up the gangplank to the ark. That two-sex model encountered hundreds of associations for the production of males or females in a new generation. Temperature, moisture, location in the uterus, location in the testes (right vs. left), direction of the wind, metabolic activity, time of the month of intercourse, and astrologically opportune time for a male (or a female) conception are just a few of the estimated 600 different theories over the past three millennia of how babies are either male or female (or hermaphrodite). Greek philosophers tackled the problem in several ways. Plato in his *Symposium* describes a banquet at which Aristophanes suggests the original humans were organized as three forms of Siamese twins. One was male–female, one was male–male, and one was female–female. They were split apart after alienating a deity, and since then they have been seeking reunion as heterosexuals, gays, and lesbians.

Aristotle and Galen Provide Models of Sex Determination from Antiquity to the Renaissance

Aristotle (384 BC–322 BC) took a more objective (but sexist) approach in his attempt to explore all extant knowledge. He thought of the female as a lesser or deformed male. Her reproductive organs were inward and males had theirs outward. He also assigned to male semen the capacity to impress form on the matter that the female provided (sometimes referred to as the female semen and sometimes as her menstrual blood). To account for male or female determination, Aristotle reverted to the assorted theories of temperature, location, and wind. Galen of Pergamon (130–210) believed sex determination was based on uterine location, and he assigned three sites to make males, three to make females, and one to make intersexes. He wrote, "Now just as mankind is the most perfect of all animals, so within mankind the man is more perfect than the woman, and the reason for the perfection is his excess of heat, for heat is Nature's primary instrument. Hence in those animals that have less of it, her workmanship is necessarily more imperfect, and so it is no wonder that the female is less perfect than the male by as much as she is colder than he."[3]

No major advances were made over the next 1500 years, but Thomas Aquinas (~1224/1226–1274) rejected the Aristotelian one-sex theory and favored a two-sex theory with the female's anatomy constructed for pregnancy and nursing. In other animals, there was more opportunity for observation. Aristotle thought of the egg in chickens as an external uterus. William Harvey adopted Aristotle's theory, but he also noted that in the deer he examined, there was no evidence of copulation leading to anything visible in the uterus. About a month later, there was a disc of material but this underwent an arrest (called diapause) for several months. This allowed the deer to time the birth of its fawns (usually twins) in the spring. With Andreas Vesalius's (1514–1564) careful depictions of gross anatomy, the reproductive structures began to be studied in finer detail. It is a surprise to learn that Galen's anatomy book used in medical schools in medieval Europe had no illustrations; his dissections were performed not on humans, but on the Barbary ape (a macaque monkey). The beautiful artistic rendering of the illustrations in Vesalius' anatomy book was performed under his supervision, most likely by Jan Stephan van Calcar, a pupil of the Venetian artist Titian. Artists had as eager an interest in anatomy as medical students had during the Renaissance. Leonardo da Vinci's notebooks show detailed drawings of human anatomy from dissections he observed and performed.

The Recognition of the Ovary as a Source of Eggs

The role of the ovary was worked out in the 17th century. The anatomical components were worked out in the late 15th and in the 16th centuries. In 1491, the first anatomical depiction of female reproductive organs was made. Until 1668 when Regnier de Graaf (1641–1673) published his findings on female anatomy and reproduction, the ovaries were known as female testicles.[4] It was de Graaf, through dissections of humans and mammals, who showed that the ovary produced eggs (although he mistook the follicle itself for the egg). He also described the morphological changes in the ovary and identified the corpus luteum on the surface of the ovary, but for more than 200 years physicians were divided on whether it was associated with yolk production or served some other function. In the 1930s, it was associated with the production of progesterone, a hormone essential for maintaining pregnancy. The oviduct was first described by Gabriel Fallopius (Gabriele Falloppio) (1523–1562). Some believed it transported or produced a female semen. Some believed it manufactured an egg. It was de Graaf who asserted it transported an egg. The actual egg in the follicle was identified by Theodor Bischoff (1807–1882) in 1843.[5] Theodor Schwann in 1834 identified the mammalian egg as a single cell.

Sperm as Parasites and as Reproductive Agents

Spermatozoa were first seen by Antonie van Leeuwenhoek in 1688, and he sent illustrations of them in his letters to the Royal Society.[6] He called them animalcules, like other infusorians and protozoa found in stagnant or pond water. It took two centuries to resolve a debate on whether these were parasites or whether they played a role in reproduction. Even when the sentiment shifted to a reproductive role in the mid-19th century, some biologists believed that several sperm were needed to bring about fertilization. Some believed that the quantity of sperm entering determined the sex of the individual. Some believed that only a single sperm entered the egg. This was resolved in animals by Otto Bütschli (1848–1920), who in 1876 described the passage of a single sperm and its eventual union with an egg nucleus. The precise physiology of that fertilization process required several additions of knowledge. Sperm shed their mid-piece and tails and formed what was called a male pronucleus. The female egg produced what was called a female pronucleus. These two pronuclei were the actual agencies of fertilization leading to a single nucleus of the next generation. At the same time, the pronuclei were associated with a specialized cell division for reproductive cells, meiosis. Meiosis was worked out with relation to the chromosomes late in the 19th century by Edouard van Beneden.[7]

Cytology Verges on Inferring Mendelism

Work on meiosis was initiated by August Weismann's suggestions, in the early 1880s, that a special division was needed for reproductive cells to reduce the chromosomal material in half. This would maintain a constancy for each generation of chromosomal material. By the time van Beneden worked out a chromosomal analysis for *Ascaris* in meiosis, it was clear that the constancy of chromosome number was significant for heredity. Cytology was heading toward heredity. This was first noted in 1891 by Hermann Henking (1858–1942) in the firebug, *Pyrrhocoris apterus*. He noted an unusual component that was either a chromosome or a nucleolus and that was unpaired in meiosis. Henking called it an X element (X for unknown). In 1898, Clarence McClung (1870–1946) noticed an unpaired chromosome in grasshoppers, but it was only present in male grasshoppers. He called it an accessory chromosome.[8] In 1900, Thomas Harrison Montgomery, Jr. (1873–1912) described the pairing of chromosomes in meiosis. In 1900, Mendel's laws of heredity were confirmed. In 1902, Walter Sutton (1877–1916), a student of McClung, and Edmund Beecher Wilson (1856–1939) associated Mendel's laws with the pairing of chromosomes in meiosis and proposed (with others, independently) a chromosome theory of heredity (i.e., that chromosomes obey Mendel's rules).[9] In 1905, Nettie Stevens (1861–1912) and, independently, Wilson described heterochromosomes (Stevens) or idiosomes (Wilson) in the organisms they studied.[10] Wilson clarified the many incoming observations and diversity by calling these chromosomes X and Y, with females having XX and males having XY chromosomes associated with sex determination. Wilson also recognized that the first hereditary trait, sex determination, could be assigned to chromosomes.

Wilson and Stevens Work Out a Theory of Sex Chromosomes

Not all species have an XY male and XX female arrangement. Wilson and his students found some of his beetles had XX female and XO male determination. Wilson also found species with multiple X and multiple Y chromosomes. Both Stevens and Wilson promoted their theory of sex determination by sex chromosomes. In studies in Great Britain, Leonard Doncaster (1877–1920) and the Reverend G.H. Raynor found the reverse association of sex with chromosomes.[11] Their pattern was designated as ZW, with ZZ being males and ZW being females. Nor do all species use a sex chromosomal basis for sex determination. In social insects such as ants, wasps, and bees, females turned out to be diploid ($2N$) and males turned out to be haploid (N). This is known as haplo-diploid sex determination.

Sex Hormones are Inferred from Studies of Freemartins in Cattle

The physiology of sex determination had a different history. As late as the 1890s, many biologists believed sex determination was associated with metabolism. There were several lines of evidence for this. Some fish would shift sexes, transforming from male to female or female to male depending on water temperature. This was occasionally seen in poultry with a hen suddenly sprouting a coxcomb and crowing in the morning and even inseminating hens. In cattle, it was known since the work of Frank Rattray Lillie (1870–1947) solved the origin of freemartins.[12] Freemartins arise when cattle give birth to twins—one male and one female. (Freemartins do not occur if the twins are of the same sex.) The male twin is normal, and the female twin—the freemartin—is genetically female, but with nonfunctioning ovaries and some male characteristics. Lillie used India ink to inject the chorions of the afterbirths, and he showed that there were anastomoses of the blood supplies of the two cattle. He assumed some hormonal influence from the male partially virilized the female twin.

Testosterone Is Identified as the Male-Determining Hormone

Testosterone, the male hormone, was isolated and purified in 1934 by Ernst Laquer (1880–1947).[13] Steroid hormones had been associated with sex determination through the work of Fred Conrad Koch (1876–1948) and Lemuel Clyde McGee (1904–1975) at the University of Chicago in 1927.[14] They used extracts of testes to restore virility to castrated chickens, pigs, and rats. Adolph Butenandt (1903–1995) and Leopold Ružička (1887–1976) in the early 1930s worked out the relation of steroid hormones to cholesterol and the structural modifications each contained.[15] The story got more complicated with the discovery of nonsteroidal hormones found in the pituitary gland, especially follicle-stimulating hormone that led to egg production and luteinizing hormone that led to corpus luteum formation. Additional nonsteroidal hormones were associated with embryonic differentiation of the internal genitals, especially through the work of Alfred Jost (1916–1991), who showed experimentally that a hormone that was not testosterone was required to destroy the embryonic tissue that would otherwise form a uterus and oviducts as well as the upper vagina.[16] This process imposed male characteristics. That hormone is known as anti-Müllerian hormone, or AMH (also known as Müllerian inhibiting substance), for its role

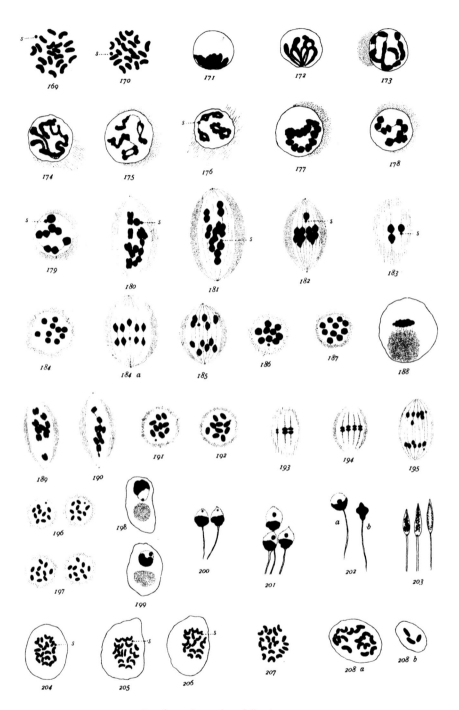

See figure legend on following page.

in causing cells of the Müllerian ducts to self-destruct. This discovery uncovered the mechanism of somatic sex differentiation and explained how male and female characteristics developed.

Hermaphrodites Revealed the Complexity of Sex Determination

As the 20th century approached the 1950s, a fairly detailed account could be made for the role of chromosomes and hormones in the formation of male or female animals, especially humans. This led to studies of babies born with ambiguous genitalia and the difficulty of assigning sex at birth when such anomalies are found. Collectively these were called hermaphrodites, but that term soon split into true hermaphrodites (those with ovarian and testicular tissue) and male pseudohermaphrodites (individuals with testes but female external or internal genitals) and female pseudohermaphrodites (individuals with ovaries but male external genitalia or with male internal genitalia). Human sex was soon divided into component events—chromosomal sex, genetic sex, gonadal sex, internal genital sex, external genital sex, pubertal sex, and psychological sex. By including the psychological, this gave biologists and physicians a possibility for interpreting the origin of gay, lesbian, bisexual, or transgendered behaviors or orientations as well as the predominant heterosexual association of most couples.[17]

The genetic aspects of sexual development include gene mutations affecting production of sex hormones and gene mutations affecting receptors on target cells that receive sex hormones. These lead usually to pseudohermaphroditic births. Some genetic events, such as nondisjunction shortly after fertilization, can lead to mosaicism, with some of the tissues having XY cells and other tissues having XO cells.[18] These can lead to ambiguous genitalia at birth. Similarly, there can be fusions of an XX preimplantation ball of cells with an XY ball of cells leading to a commingling of these cells in a common blastocyst and a pregnancy that

---

Nettie Stevens, who was mentored by E.B. Wilson and T.H. Morgan at Bryn Mawr, studied what were called accessory chromosomes that had been studied and disputed since Clarence McClung found them in 1898. Independently, she and Wilson found some species with one accessory chromosome present alone or as a pair of homologous chromosomes. Stevens used the mealworm beetle *Tenebrio molitor* and Wilson used reduviid bugs. Independently, they found that the accessory chromosome (called an X element by McClung) was widely present in sexually reproducing species. Wilson used the letters X and Y to present the accessory chromosomes. In this plate from Stevens' work, the item 184 a shows a smaller chromosome (later called the Y by Wilson) separating from the X. Both Wilson and Stevens reported XX female and XY male systems and XX female and XO male systems, in which O is the absence of an accessory chromosome. In both systems, the autosomes (chromosomes other than the sex chromosomes) are represented as paired components.

results in a chimera or infant with mixed gonadal development and mixed internal and external sexual differentiation. Those were the "true" hermaphrodites of the 1950s to 1990s. In the past 20 years, those with ambiguous genital development have organized to press for a change in terminology. They describe their conditions as DSDs, or differences in sexual development (but many physicians think of this as disorders of sexual development).[19] Patients think in terms of self-image. Physicians think in terms of disorders, anomalies, and diseases.

Genes Involved in Sex Determination Are Isolated Late in the 20th Century

There are other genes involved in sex determination, a very large field now in both biology and medicine. One of the more dramatic findings in the 1990s was a gene on the Y chromosome associated with male determination. No such gene is found on the Y chromosome of fruit flies, where a different mechanism is involved. But in mammals there is a gene called SRY for sex-regulating region of the Y.[20] If a zygote is XY with SRY present, male development occurs. If a mutation or deletion of SRY occurs in the Y, such XY zygotes produce XY females with vagina, labia, clitoris, and internal uterus and oviducts. They lack gonads. At puberty, they require female hormones by medical prescription. If the SRY gene gets inserted into an X chromosome, the modified XX individual may be born as a male with penis and scrotum and testes, but the testes are sterile because most of the genes for spermatogenesis are found on the Y chromosome.

There is less convincing evidence of genetic mechanisms involved in sexual orientation. Behavior is immersed in culture and acquired knowledge and habits

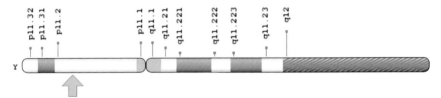

The Y chromosome in humans contains a gene called *SRY*, location at line 1 below the illustration shown above, at Yp11.2 (see arrow). Embryos with the gene become males with external and internal male genitalia. The X chromosome does not have the *SRY* gene. Females have two X chromosomes and males have only one. Yet most of their genes produce the same amount of X chromosome products. Muller called this dosage compensation in 1932 and worked out the genes in fruit flies that made it possible to equalize doses of male and female X chromosome genes. In humans, some 35 years later, Mary F. Lyon (1925–2014) found XX females carry out dosage compensation by inactivating one of the two X chromosomes. Females are thus mosaics with half their cells expressing paternal and the other half expressing the maternal X. Note that the same term is used for two different mechanisms bringing about the same desired outcome.

as well as uncritically accepted worldviews (e.g., patriotism, religion of upbringing, and ethnic sentiment). This makes it hard to tease out what is learned and what is innate. There is one school of evolutionary psychologists who favor innate construction of male or female identity or for sexual orientation. Even those who favor this are divided among those who favor a genetic innate basis for orientation and those who favor hormonal stresses during pregnancy (especially during late organogenesis when sexual differentiation occurs or during the development of the brain in the fetal stage). Opposed to those invoking a biological basis for sexual orientation are those with a strong religious belief that sexual orientation is mandated to be heterosexual and departure is a consequence of choice and hence a moral issue subject to religious approval or disapproval. Also opposed to a biological basis for sexual behavior are those, especially in the social sciences, who see all behavior as constructed by social influences. No doubt, as the Human Genome Project identifies dozens or hundreds of genes involved with sex determination, some of these issues will be settled.

The Origin of Sex Is Related to Genetic Recombination

What I have presented is a very broad overview of an immense field of knowledge that today would be considered an interdisciplinary effort in understanding sex

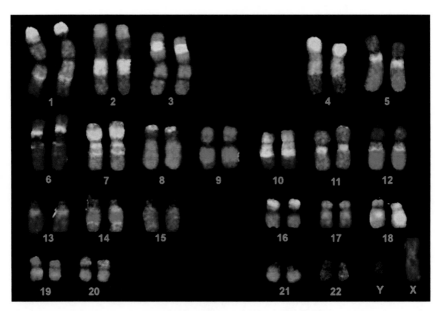

The human chromosomes as seen by fluorescence in situ hybridization, or FISH. The X and Y human chromosomes are on the *lower right*.

determination. It is quite varied among the millions of species and sexuality is found among bacteria and all phyla. Even viruses have something similar to sexuality when viruses of different strains infect a common host and exchange genes. For the biologist, the key to interpretation of sexuality is the way genes can be recombined in a population among members of a species. This is what meiosis brings about at the genetic level as Mendelian recombination and segregation of character traits. For sex determination, the fields of endocrinology, cytology, genetics, anatomy, and embryology are essential for understanding that sexual development to birth. Psychology gets thrown in when gender and orientation are added later in the life cycle. Having written a book about the history of sex determination, I believe I have left out 90% of what I covered there

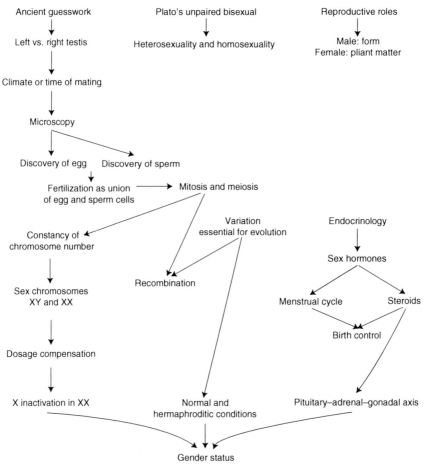

See figure legend on following page.

and it makes me appreciate how limited the historian's quest is in trying to reconstruct the past by selecting a small portion of it for analysis. Despite that inherent bias of omission, I see no major paradigm shifts in the way sex determination moved from guesswork, to anatomy, to function of the genitals, to microscopic identification of sperm and egg, to microscopic analysis of meiosis and its role in gamete formation, to sex chromosome identification, to the chromosome theory of heredity, and to the endocrinological and genetic basis of embryological development of the sexual organ system.

At present, the molecular biology of sex determination with numerous genes and their biochemical roles in cells is unfolding.[21] From the point of view of society, the knowledge required to make sense of these findings or to critically judge them becomes remote for those not in the field or cognate fields. Because of the importance of sexuality in our self-identity and the way society treats deviations from cultural norms, this is a major unresolved problem that society has for the most part not assimilated.

In antiquity, scholars believed females arose from the left testis and males from the right, or they believed that the time of copulation, or temperature at the time of copulation, determined male or female conception. In Plato's *Symposium*, Plato argues sex was originally a paired male and female. They were cut into separate sexes and later tried to get together, mostly as male–female pairs, but homosexuality resulted from male–male or female–female halves combining. Aristotle believed males provided form to the female's matter (sometimes believed to be menstrual blood). After the microscope was introduced, sperm and eggs were identified, but it was not until the 1870s that a single sperm nucleus was followed to its union with a female egg nucleus to produce a zygote. The study of mitosis and meiosis led to the theory of the constancy of the chromosome number (diploid) except for gametes (haploid). Weismann argued that meiosis provided genetic recombination and that was the primary biological function of meiosis. That variation in gametes accounted for Darwinian variations in populations. Hormones were first identified in the late 19th century and sex hormones (steroids and pituitary peptide hormones) were identified in the early 20th century. Wilson and others found sex chromosomes, and he designated XY as male and XX as female. Morgan worked out X-linked inheritance. Genes for sex determination were isolated by Morgan's fly lab. The menstrual cycle was worked out and allowed birth control to become effective, especially when hormones could be used to prevent ovulation ("the pill"). Dosage compensation (why XX and XY individuals have the same gene product content) was worked out in fruit flies and it differs from mammals, which use X inactivation. The study of sex chromosome nondisjunction—XXY, XO, XXX, XYY, and mosaics X/XY or chimeras XX/XY—and gene mutations involving the pituitary–adrenal–gonadal axis led to classification of hermaphroditic and pseudohermaphroditic conditions (now called intersexual conditions). There are many ways animals and plants have worked out sexual differences. The sex chromosome method is the most common among multicellular animals. Gender is largely cultural in human social practices.

References and Notes

1. Lillie FR. 1919. *Problems of fertilization*, p. 14. University of Chicago Press, Chicago.

2. Wilson EB. 1910. The chromosomes in relation to the determination of sex. *Science* **4**: 570–592.

3. McCartney E. 1922. Sex determination in antiquity. *Am J Philos* **43**: 62–70. Also see Mittwoch U. 2005. Sex determination in mythology and history. *Arq Bras Endocrinol Metab* **49**: 7–13.

4. De Graaf R. 1701. *Nouvelles Déscouvertes sur les parties de l'homme et de la femme, qui servent à la generation, avec la déffense des parties génitales, contre les sentiments de quelques anatomistes, un traité de pucelage, du pancreas, de l'usage du siphon et des clystères*. In Harvard University Library, Hollis Library, Cambridge, MA.

5. Bischoff T. 1842. *Entwicklungsgeschichte der Säugetiere und des Menschen*. Voss, Leipzig, Germany.

6. Leeuwenhoek A. 1683. An abstract of a letter from Mr. Anthony Leeuwenhoek of Delft about generation by an animalcule of the male seed. Animals in the seed of a frog. Some other observables in the parts of a frog. Digestion and the motion of the blood in a beavr [*sic*]. *Philos Trans Royal Soc* **13**: 347–355.

7. Hamoir G. 1992. The discovery of meiosis by E. van Beneden: A breakthrough in the morphological phase of heredity. *Intl J Dev Biol* **36**: 9–15.

8. McClung C. 1902. The accessory chromosome—Sex determinant? *Biol Bull* **3**: 43–84.

9. Sutton W. 1903. The chromosomes in heredity. *Biol Bull* **4**: 231–251.

10. Wilson EB. 1905. Studies on chromosomes. I: The behavior of the idiochromosomes in *Hemiptera*. *J Exp Zool* 2: 371–405. See also Stevens NM. 1905. *Studies in spermatogenesis with especial reference to the accessory chromosome*. Publication number 36. Carnegie Institution, Washington, DC.

11. Doncaster L, Raynor GH. 1906. Breeding experiments with *Lepidoptera*. *Proc Zool Soc London*, Part **1**: 125–132.

12. Lillie FR. 1916. Theory of the freemartin. *Science* **43**: 611–613.

13. David KA, Dingemanse E, Freud J, Laquer E. 1935. Über krystallinisches männliches Hormon aus Hoden (Testosteron), wirksamer als aus Harn oder aus Cholesterin bereitetes Androsteron. [On crystalline male hormone from testicles (testosterone) effective as from urine or from cholesterol]. *Hoppe-Seylers Z Physiol Chem* **233**: 281–282.

14. McGee LC. 1927. The effects of a lipoid fraction of bull testicles in capons. *Proc Inst Med Chicago*. Ph.D. dissertation in Fred Koch's laboratory, University of Chicago, Chicago.

15. Butenandt A. 1929. Über die chemische Untersuchung der Sexualhormone. *Z Agew Chem* **44**: 905–908.

16. Jost A. 1953. Problems of fetal endocrinology: The gonadal and hypophyseal hormones. *Rec Prog Horm Res* **8**: 379–418.

17. Carlson EA. 2012. *The seven sexes: Biology of sex determination*. Indiana University Press, Bloomington.

18. Mittwoch U. 1967. *The sex chromosomes*. Academic Press, New York.

19. Hughes IA, et al. 2006. Consensus statement on management of intersex disorders. *Arch Dis Child* **91**: 554–562.

20. Sinclair AH, Berta P, Palmer M, Hawkins JR, Griffiths BL, Smith MJ, Foster JW, Frischauf AM, Lovell-Badge R, Goodfellow PN. 1990. A gene from the human sex determining region encodes a protein with homology to a conserved DNA-binding motif. *Nature* **396**: 240–244.

21. Eggers S, Sinclair A. 2012. Mammalian sex determination—Insights from humans and mice. *Chromosome Res* **201**: 215–238.

CHAPTER 8

Genotype and Phenotype Relations: From Variations to Genetic Modifiers to Epigenetics

Creating new breeds by hybridization, Mendelian independent assortment and complex multigene traits, epistasis, recombination associated with crossing-over. Constancy of the gene in repeated isolation in the dark, constancy of the gene in multigenerational heterozygous state. Genotype versus phenotype in inbred lines, environmental shifts of expression of genes, "anastomoses" of gene interactions in beaded and truncate. Chief genes and modifiers, temperature-sensitive mutations, conditional mutations. Heterogeneity in human disorders, penetrance and expressivity, polymorphism and evolution.

Before the 20th century, the terms "genotype" and "phenotype" did not exist. In popular imagery, children were sometimes described as "chips off the old block," but at the same time Gilbert and Sullivan warned, "Things are seldom what they seem." The former lends itself to the prevailing practice of animal breeders and horticulturists mating like with like to produce a fixed breed. But every breeder ran into an occasional "atavism," or reversion to an ancestral type. The quote from *HMS Pinafore* suggests that there may be a difference between what we see and what is actually present. Breeders could also testify to the frustration of mating a prize specimen and finding it did not transmit some desired traits.

There was also a disturbing aspect in relating appearance to intrinsic worth. The introduction of phrenology by Franz Joseph Gall (1758–1828) in 1796 suggested that the slight bumps and irregularities of the skull could predict talents, personality, and even criminality. Phrenology was believed to be a serious science for several generations before it fell into disfavor as a pseudoscience. A similar, but older system going back to Greek antiquity argued that facial appearance, or physiognomy, was a clue to personality. Variants also existed for palmistry, with the lines or creases in a person's hands as a map to the person's status, past, and future. Not very different, but sounding less credulous, was the belief that race, ethnicity, or class was associated with innate behavioral traits. Carl Linnaeus (1707–1778) believed each major race had a characteristic set of personality traits, most of them not very flattering, but he resisted the assignment of races to new species or subspecies of humans. They were all *Homo sapiens* in his fidelity to the physical features he studied for classification. Later anthropologists like Joseph Arthur, Comte de Gobineau (1816–1882), in *An Essay on the Inequality of the Human Races* (1853–1855) would create a scientific racism in which skin color was the major guide to classifying human abilities and temperament. Those phenotype–genotype associations still prevail wherever ethnic rivalries, racial discrimination, and class warfare exist. Blue bloods flatter themselves. The elite assign their success to their good stock, and for centuries, a British gentleman would be assumed to be of upright character by birthright. Those who were identified as social failures were looked on as suffering from a long-standing inbred degeneracy that they would pass to their future descendants unless the traits were so debilitating they led to extinction of the family line.

The Nature of Hybrids Raised Questions about Heredity

Complicating the issue of appearance and innate qualities were the different views on hybrids. Some hybrids were seen in derogatory terms. Marrying a person of another race led to terms such as "mulatto" and "quadroon" for black-and-white offspring. The term "mestizo" was applied to the progeny of Hispanics and South American Indians. The term "zambo" was applied to progeny of Native Indians with Africans. Those terms were dropped after the revolutions in South America initiated by Simón Bolívar (1783–1830). He defined anyone who had any Hispanic ancestry as a Latin American. For Bolívar the Liberator, his term was inclusive.[1] For the slave owners and exploiters, their terms were exclusive. Hybrids, bastards, and those of poor stock were seen as inferior and they were often barred from their rights as citizens.

With natural species, there was considerable debate on the nature of hybrids. Some, like Linnaeus, praised hybrids as the sources of new varieties and species. Some saw hybrids as weakening each participating race or breed. In general, zoologists and botanists used the term "heredity" to apply to the characteristics that define a species. Variations described those subtle departures within a species that distinguished individuals. Rogues and sports were undesirable variations that breeders generally kept from breeding. They usually arose in cultivated plants and animals. Varieties were established lines either in nature or through the breeder's art. Species had fixed Linnaean characteristics, which they shared in such a way that excluded members of any other species. The terms began to blur and conflict because breeders were beginning to detect components of what would later be attributed to Mendelian analysis. They also blurred because they were picked up by Charles Darwin and his supporters as the theory of evolution by natural selection was introduced in 1858.[2]

A third aspect entered into the fray when a popular belief in the inheritance of acquired characteristics was shifted by Jean-Baptiste Lamarck (1744–1829) to the new field of biology in 1802. His theory of evolution by environmental modification of traits in response to need was met with split response. Georges Cuvier (1769–1832) hated it and thought Lamarck unworthy of honor as a scientist. Scientists of Darwin's generation agreed with the lack of evidence to support his theory. Darwin did not want to make that mistake, and he downplayed Lamarckism; but he reverted back to it when he was forced to seek a hereditary basis for natural selection and the origin of the variations on which it acted. Numerous studies of hybrids in the 19th century were tried. The studies of Josef Gottlieb Kölreuter (1733–1806) showed that if two varieties of maize were crossed, the offspring showed one of the parent's traits. But if the hybrid were self-pollinated, the cobs of the second generation showed rows and isolated distribution of the two different traits. Later, Mendel would describe this as a segregation of contrasting traits. To Kölreuter, the interpretation was one of incompatibility. The two strains were at war with each other and the hybrid was inherently unstable.[3] The term "hybrid," coined from the Greek word "hubris," was chosen to reflect that accepted belief that the formation of hybrids was an abomination or unnatural act. Although Darwin did not share such a supernatural interpretation, he performed crosses to see if the entire plant evolved rather than focusing on specific isolated traits. He chose highly variable traits like height rather than strikingly distinct traits. Height was distributed in a bell-shaped curve as Galton later showed. Whereas Darwin drifted off to his theory of pangenesis and some vague circulation of changing units (gemmules) in the organism, Galton looked for mathematical rigor in describing traits in populations. He hoped this might show how traits evolve.

Mendel's Laws Provided an Explanation for Hybrids

By the end of the 19th century, there was the forgotten or ignored work of Gregor Mendel, which showed fixed ratios for discontinuous traits. Even if Darwin had been shown Mendel's work, there is little likelihood it would appeal to his view that evolutionary traits are built in subtle increments from highly variable units of heredity. In contention were the Darwinists of the biometric school looking for a mathematical solution to speciation by studying curves of highly variable traits. Also in contention was the proposal by Bateson that discontinuous variation was the source of evolutionary change, especially repetition of parts (meristic variation) and homeotic mutants in which organs shifted to new locations.[4] The study of hybrids dramatically changed for zoologists and botanists when Mendel's work was rediscovered or confirmed and extended in 1900. Hugo de Vries used more than a dozen different plants to confirm Mendelism. The new generation of breeders, especially Bateson and his students, found new phenomena, and Bateson was encouraged in 1906 to name a new branch of physiology—genetics.

The new phenomena included the modified ratios of 15:1, 9:7, or 9:4:3, all modifications of Mendel's second law of independent assortment represented by a 9:3:3:1 ratio. The interaction of two factors in producing a color, for example, was called epistasis. When there was no dominance involved and a trait was classified as quantitative, as in Herman Nilsson-Ehle's (1873–1949) cereal grain coat color, the range ran from red to dark pink to pink to light pink to blond or white. If there were two pairs of factors involved, the bar graph was in a ratio of 1:4:6:4:1 for the color of the cereal grains. To some degree, the genetic constitution that Bateson called its genotype could be inferred from the genotype. Hybridization confounded that possibility in some of these categories. The parents were AA BB (red) and aa bb (white). The F_1 offspring were Aa Bb (pink), the F_2 from the cross of two hybrids yielded the 1 (red or AA BB) to 4 (dark pink or AA Bb or Aa BB) to 6 (pink from AA bb or aa BB or AaBb) to 4 (light pink or aa Bb or Aa bb) to 1 (white aa bb). For Mendel, as for Bateson and Nilsson-Ehle, the genotype could predict the phenotype, but the phenotype could only predict limits or possibilities to the number of genotypes for a given phenotype.

Quantitative Traits and Polygenic Traits Were Represented Mathematically

If the number of factors increased to seven or more, it was difficult to distinguish a bar graph representing subtle shades from red to white, and such distributions began to look like normal curves. When the normal curve was obtained and no gradations were possible, as in Wilhelm Johannsen's study of inbred beans selected for size, the original bell-shaped curve started with a distinct mean. By choosing the largest bean, the mean shifted each generation for about 10 or 12 generations and then it no longer shifted. Johannsen drew the conclusion that the variation of the purebred-selected nonshifting line was environmental and represented the range of a homozygous genotype, all sources of hybridization having been eliminated by this rigorous selection. Johannsen worked out his bean size experiments between 1902 and 1910.[5] He argued that both genetic variation and environmental variation were possible for a trait, and he had showed this although the number of genes involved was unknown. Johannsen introduced the idea of the gene in 1909, but he could not assign any attributes to it. He rejected the term used by Bateson—"unit character"—because it struck him as confusing—a character trait with a unit of heredity. It was better to have a vague term—"gene"—than to imply the gene itself was the character or directly made that character. That was a very valuable insight for the young science of genetics, but the lesson needed many repetitions before it became a fixture of classical genetics.

Hybridization and Hybrid Vigor

In 1908, George Harrison Shull (1874–1954) published "The composition of a field of corn."[6] He did his work at the Carnegie Institution's Station for Experimental Evolution at Cold Spring Harbor, New York, and showed that inbred lines of maize tended to be low yielding in bushels of corn produced, but cross-bred lines of corn produced more uniform and more vigorous

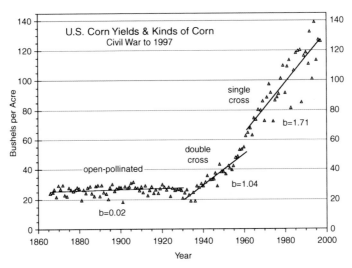

Hybrid corn contributed to the wealth of American farmers in the mid-20th century and continues to be the preferred method of farming. The increased yield and quality of hybrid corn was given the name heterosis by Shull, who did not try to interpret it. Two models contended. The dominance group claimed that crossing inbred lines restores heterozygosity and reduces harmful inbred recessive gene mutations. The overdominance group claimed that heterozygosity itself may confer benefits neither homozygous gene could produce. Some hybrids, like mules, are stronger and versatile farm animals, but they are sterile. It is not overall fitness but trait-by-trait comparison that reveals the different expressions of genes when heterozygous or homozygous.

plants with increased bushels of corn. In 1909, he described how he would first establish inbred lines and then cross them to make the hybrid generation of seed.[6] His work attracted the attention of Edward Murray East (1879–1938), who confirmed Shull's findings. East and his student Donald Forsha Jones (1890–1963) worked out a "double cross" or "four-way" cross in which inbred strains A and B produced hybrid 1 and inbred strains C and D produced hybrid 2. By crossing hybrid 1 × hybrid 2, East and Jones got even more yield per acre.[6] The commercial use of hybrid corn started slowly, but got a boost when Henry A. Wallace (1888–1965) sold hybrid seed he produced to neighboring farmers.[6] Wallace later became Secretary of Agriculture and then Secretary of Commerce for President Franklin Roosevelt; he also served as Vice President in the Roosevelt administration. The Dust Bowl drought of 1933–1935 helped Wallace's business because the hybrid seeds survived and produced more corn than the traditional family field corn. Within 10 years, almost all farmers in the Corn Belt states were using hybrid seed for their corn crops.

How the Fly Lab Interpreted Genotype/Phenotype Relations

Challenging Johannsen's interpretation of quantitative traits were critics of T.H. Morgan and his school. William Castle was perhaps the most outspoken in his view that when a trait's phenotype varied, it was likely to reflect a variable fluctuation of the gene. Castle worked with mammals. Morgan and his students worked with flies. Castle chose variable traits like hooded rats or spotted rabbits.[7] Although these Mendelized and thus were associated with a single gene that was mutant, the phenotypes of the animals that expressed the mutant gene were highly variable. Castle showed they could vary, expressing a hooded pattern with very little white fur to offspring with just a minor hooding and an almost all white body. Why did they vary? To Castle, heavily influenced by the idea of Darwinian fluctuations, this was evidence of a gene that varied. Why did it vary? To Castle, one cause was contamination by prior breeding with a normal allele and this resulted in an admixture to the trait. If the gene fluctuated so did the character expressed. But Morgan's group disagreed, especially young Hermann J. Muller, who in 1914 accused Castle of "a spirit of mysticism" in invoking such a proposal without evidence. Rather, Muller claimed (without abundant evidence) that variable traits were subject to both genetic and environmental modifiers. Muller had just started working on beaded wings and truncate wings. Morgan had found them and he had difficulty getting them to Mendelize. Muller asked if he could give them a try.

Muller devised a system to suction individual male flies by using a modified pipette and then mating them to different females. The stocks he designed carried genetically different markers so he could follow them and their progeny. He identified beaded and truncate as "chief genes." Without them, the mutant trait would not be expressed. He identified, named, and then mapped two types of modifier genes. Some were intensifiers that exaggerated the mutant expression of the chief gene. Some were diminishers that normalized the expression of the chief gene. Muller could combine different intensifiers or diminishers, and when the chief gene was present he could predict how many of each degree of mutant expression would be found and how many would show normal wings. For Muller, it was stunning evidence that discontinuous hereditary units—genes—could produce a range of expression formerly attributed to Darwinian fluctuating variations. Muller extended this to natural selection and showed that classical genetics was compatible with Darwinian evolution. He promoted this

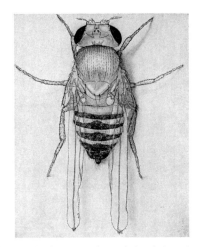

In 1918, Muller published his analysis of beaded wings. He showed that it is a dominant mutation for wing expression, but it could not be rendered homozygous because it also acted as a recessive lethal. The scalloped wing is shown in this drawing with average and extreme expression. There was a range also from mild scalloping to this extreme form. Muller found that there was a chief gene necessary for expressing beaded wings. He also isolated two categories of modifiers, intensifiers of the mutant expression and diminishers that made the wing look normal. He mapped those modifiers and used the combinations along with an environmental factor, temperature, to predict the percent of wing modifications along the spectrum of wing scalloping. The analysis solved problems in genetics at that time, including how fruit fly mutations could account for Darwinian gradual selection for character traits, how lethal mutations could create perpetual hybrids, and how these could accumulate mutations over generations, as often occurs in evolution.

interpretation in papers between 1917 and 1921.[8,9] Soon, mathematical population genetics would emerge through the work of R.A. Fisher (1890–1962), J.B.S. Haldane (1892–1964), and Sewall Wright (1889–1988) to put in mathematical terms the number of genes and the relation of genes to traits and churn out their incidence in the population under different environmental effects. This two-pronged approach of experimental analytic genetics teasing apart the components of variation into genetic and environmental components and the mathematical construction of population genetics shifted classical genetics into Darwinian evolution by natural selection.

Human Genetic Traits Were Described as Having Penetrance and Expressivity

Muller's work on truncate and beaded wings led Oskar Vogt (1870–1959) to propose a similar mechanism for the variability of expression for human genetic disorders.[10] Sometimes a dominant disorder would not express (like a gene for retinoblastoma that causes cancer in the eyes of children). These were called "penetrance" mutations. A mutant with high penetrance usually expressed when the gene was present, as in achondroplastic dwarfism. A gene with low penetrance might express in only 50% or less of the fertilized eggs getting that mutant gene. Muller's work suggested that genetic or environmental modifiers might influence the penetrance of the chief genes. A child with achondroplastic dwarfism could also have curvature of the spine (lordosis) or a prominent forehead or a narrow foramen magnum through which the spinal cord passes and that, when too small, could lead to hydrocephalic development. These different expressions were designated as a gene's "expressivity." High expressivity meant most or all of the possible expressions of that gene appear in the child. Low expressivity meant that only one or a few of the traits varied expressions showed. N.V. Timoféef-Ressovsky, who worked in Oskar Vogt's institute in Berlin, popularized Vogt's new terms for both fruit fly genetics and human genetics.

The causes of reduced penetrance and variable expressivity sometimes involve genetic modifiers, sometimes involve environmental factors, and sometimes involve the unusual structure of the gene itself, which might have numerous repeated nucleotide sequences. These often lead to mispairings and the formation of larger sets of the repeated elements or lower sets. A major class of these defects involves "trinucleotide repeats."[11] Usually the larger sets lead to high penetrance and high expressivity, as in Huntington disease. When geneticists shifted from fruit flies and maize as classical organisms for genetic research, they filled the field of viral and bacterial genetics because these organisms were easier to analyze in biochemical or molecular contexts. One category of mutations responded to temperature. At a higher temperature, they expressed the mutant trait but at a lower temperature they expressed the normal trait. A fruit fly geneticist might have believed them to be haploid instances showing reduced penetrance. Microbial geneticists instead called them conditional mutations or temperature-sensitive mutations. Muller and Altenburg made use of this conditional response of the chief gene truncate wings to temperature and showed that at higher temperatures the mutant expression is favored. Another category of mutations usually results in total failure of normal expression because a triplet for an amino acid is converted to a stop signal, resulting in a shortened protein that is usually nonfunctional. In viruses, these are called amber mutations.[12] The latter is an example of idiosyncratic naming of genes. They exist for all organisms and sometimes reflect wit, a private joke, or some testimonial to an incident bringing the mutant trait to the attention of the person who found it. The geneticist who found those chain-terminating mutations in phage had argued about such a possibility with his colleague. The discoverer told his friend that if he found such mutations he would name them for him. His friend's name was Bernstein. This was somewhat long for a mutation's name, but Bernstein is the German for amber and amber mutations they became.

Human Skin Color Is a Quantitative Trait

The relation of genotype to phenotype is a significant aspect of genetics. When the analysis is performed, the relation can be clarified. That is possible down to the molecular level for many human traits. That is true for single-gene disorders in humans that presently involve several thousand instances of autosomal recessive, X-linked, or autosomal dominant conditions. But it is more difficult to do when several genes are involved in the expression of a trait. This includes the possibility of a chief gene leading to cancer or quantitative traits in which several genes are known to be involved, like skin color determination in humans. Charles Davenport (1866–1944) and his colleagues studied this in 1913 in Jamaica and concluded that the model proposed by Nilsson-Ehle was closest to what they found for human skin color among interracial families.[13] The familiar 1:4:6:4:1 ratio from black skin (similar to a West African) to white skin (similar to someone from Norway) would include dark brown, brown, and light brown individuals. Charles Davenport used a "color top" that he obtained from a toy company and matched the color of the inside of the upper arm to the spinning wedges of yellow, white, red, and black on the top that blurred into the appropriate matching shade of the skin. Davenport's model was useful in dispelling racial myths about the inheritance of skin color and other traits from distant ancestors who may have had those genes and which faded from family memory over the generations. It showed how two individuals of light brown skin color could have children darker than them. It also showed how two heterozygous brown-skinned individuals could have several children expressing a range of colors from white to black. He ruled out the possibility of a "black baby" from a white couple, a favorite myth in the American South to discourage racial mixture. In the 19th century, that child would have been called an atavism. In the 20th century after Davenport's analysis, it would have been called a racial myth that was virtually impossible to occur.

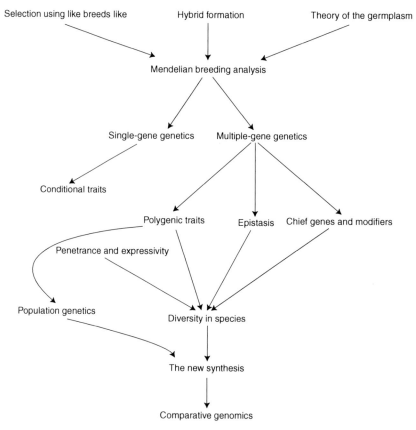

See figure legend on following page.

How Should Science Interpret Anomalous Results?

Science is most effective when it can explain anomalous results. These occur in Kuhn's normal science and they occur in incremental models of progress as well. From the perspective of 19th century science, the type of detailed analysis in the 20th century that is described here was unimaginable. Experiment by experiment, the apparent contradictions to expectation arose and dissipated as modifier genes, environmental modification, and other features of gene–character associations were worked out. The shift from folk heredity to a more precise identification of components in each new finding of variability and complexity of character traits was incremental. What we call "classical genetics" emerged from breeding analysis, epistasis, chief genes and modifiers, quantitative analysis, cytogenetics, population genetics, and Darwinian selection. Gene character relations occur through many ways. Each anomaly is explained not as a series of mini-paradigm shifts, but by the numerous tools, experiments, and findings as a field expands.

In agricultural practice since antiquity, the prevailing belief was like breeds like. It usually works. Exceptions included "sports" or new mutations and reversions to less desired forms. Sometimes novelties were beneficial (like extra rows on an ear of corn). A second method used by breeders involved hybridization. Horticulture uses this method to produce thousands of new varieties of ornamental flowering plants. Both methods assume that traits are largely inherited and the genes involved are not modified by the environment. That assumption was reinforced by Weismann's experiments and writings promoting the "theory of the germ-plasm." In it, Weismann claimed acquired changes in body tissues and organs are not passed on through the reproductive tissue, which is restricted to the ovaries or testes. Mendelian analysis showed how traits are passed on with predictable outcomes (Mendel's ratios). In single-gene genetics, there are conditional mutations that are temperature sensitive. Many single-gene traits have multiple tissue or organ effects. Not all possible expressions occur in a single individual. This is called penetrance for showing the abnormal mutant trait and expressivity for the number of associated expressions that occur in the individual. Many traits take two or more genes to achieve expression. Bateson identified epistasis or modified Mendelian ratios (9:7 or 15:1 or 9:4:3 instead of 9:3:3:1). Muller showed that a chief gene may have genetic or environmental modifiers to produce a range of intensities of expression of a trait. Sometimes a variety of recessive alleles are present in a population and the environment can select for one or a few of the variants present. This is called polymorphism. All of these lead to diversity in a population on which natural selection acts. This led to the "new synthesis" in which classical genetics, population genetics, and cytogenetics were merged with paleontology, comparative anatomy, and geographical distribution to interpret evolutionary relations of populations. In more recent times, the recognition of DNA as the genetic material and the ability to sequence it has created a field of comparative genomics that allows evolution to be followed by the mutational changes among a population around the world or even from extinct or mummified tissue.

THE

AMERICAN NATURALIST

VOL. XLIV   *November, 1910*   No. 527

HEREDITY OF SKIN PIGMENTATION IN MAN

GERTRUDE C. DAVENPORT AND CHARLES B. DAVENPORT

CARNEGIE INSTITUTION OF WASHINGTON, STATION FOR EXPERIMENTAL
EVOLUTION, COLD SPRING HARBOR, N. Y.

641

See figure legend on following page.

Charles Davenport began his scientific career as an engineer but switched to studying the life sciences. He embraced both quantitative and experimental approaches to heredity and published widely at the turn of the 20th century. He was also an ambitious administrator and was selected to head the Carnegie Institution of Washington Station at Cold Spring Harbor, New York. There he set up three units. One was for basic research in genetics. The second was a teaching center for high school and elementary school teachers where they would learn science to use in their classes. The third was a Eugenics Record Office that would apply new findings in genetics to human heredity. The paper on human skin color that he wrote with his wife shows his methodical way of studying human skin color in tones of Caucasians (whites) and in people then designated as negroes (item D in table of contents shown). He used Nilsson-Ehle's polygenic quantitative analysis in interpreting human skin color as involving four additive genes (black), with those with three as dark brown, those with two as brown, and those with one as light brown. Those with none of the four major skin-color melanizing genes he called blond. In the last part of the Davenports' analysis, they discuss albinism as a different type of defect involving melanin synthesis failure with resulting damage to vision as well as white hair and very pale skin color (even in African albino families). Davenport was less critical as a scientist in taking almost every human behavior trait as driven by gene mutations, especially social failure or success. This emphasis on eugenics, based on social traits assumed to be inherited, tarnished his reputation and attracted bigots who believed Catholics, Jews, Eastern Europeans, and Southern Europeans should be restricted from immigration and limited in their reproduction.

References and Notes

1. Mörner M. 1967. *Race mixture in the history of Latin America*. Little, Brown, and Company, New York.

2. Carlson EA. 2011. *Mutation: The history of an idea from Darwin to genomics*. Cold Spring Harbor Laboratory Press, Cold Spring Harbor, NY.

3. Olby R. 1966. *Origins of Mendelism*. Constable, London.

4. Schwartz J. 2010. *In pursuit of the gene*. Harvard University Press, Cambridge, MA.

5. Johannsen W. 1909. *Elemente der Exakten Erblichkeitslehre*. Fischer, Jena, Germany.

6. Shull GH. 1908. The composition of a field of corn. *Am Breeders Assoc Rep* 4: 296–301; Shull GH. 1909. A pure line method of corn breeding. *Am Breeders Assoc Rep* 5: 31–59; East EM, Jones DF. 1919. *Inbreeding and outbreeding: Theoretical genetics and sociological significance*. Lippincott, Philadelphia; Crow JF. 1998. The beginning of hybrid maize. *Genetics* 148: 923–928.

7. Castle W. 1911. *Heredity in relation to evolution and animal breeding*. Appleton, New York.

8. Muller HJ. 1914. The bearing of the selection experiments of Castle and Phillips on the variability of genes. *Am Natur* 48: 567–576.

9. Muller HJ. 1918. Genetic variability, twin hybrids, and constant hybrids, in a case of balanced lethal factors. *Genetics* 3: 422–499. See also, Altenburg E, Muller HJ. 1920. The genetic basis of truncate wing—An inconstant and variable character in *Drosophila*. *Genetics* 5: 1–59.

10. Stern C. 1960. O. Vogt and the terms "penetrance" and "expressivity." Letter to editor. *Am J Hum Gen* 12: 141.

11. Kremer EJ, Pritchard M, Lynch M, Yu S, Holman K, Baker E, Warren ST, Schlessinger D, Sutherland GR, Richards RI. 1991. Mapping of DNA instability at the fragile X to a trinucleotide repeat sequence p(CCG)n. *Science* **252:** 1711–1714.

12. Crow J, Dove W. 1995. The amber mutants of phage T4. *Genetics* **141:** 439–442.

13. Davenport GC, Davenport CB. 1910. Heredity of skin pigmentation in man. *Am Natur* **44:** 641–672, 705–731.

Microbial Life: From Invisible Spores to Germs and Prokaryotic Organisms

Fracastoro's spore hypothesis, microscopy and animalcules, Pasteur and spoilage microbes, Pasteur and germ theory, Koch's postulates. Unicellular bacteria, protozoa, fungi, and algae. Bacterial division (amitosis), bacterial anatomy and electron microscopy. Fungi and biochemical pathways. Bacteria and gene structure, bacterial and viral relations through lysogeny, microbial reproduction. Genetic basis of pathology, molecular basis of fermentation, microbes as factories for human products, synthetic microbes.

Before there were microscopes, no one knew that there was a field of biology that would be called microbiology. But Girolamo Fracastoro, who practiced medicine in Verona in the early 16th century, did speculate that such a possibility existed for contagious disease, and he called these agents spores.[1] Fracastoro did not see them; he also did not know if these were spores as a type of dust of mineral nature or actual living things. He applied his theory to syphilis, which he named; he assumed he would find a treatment for it that did not rely on the imbalance theory of four humors that went back to Galen. Fracastoro was one of the new breed of Renaissance physicians that would later include the Swiss physician (and alchemist and occultist) Paracelsus (born Philippus Aureolus Theophrastus Bombastus von Hohenheim, 1493–1541) who sought empirical medicine as a better way to understand disease and health. They also tried different substances to block the progress of disease. Fracastoro tried mercury. It worked in some of his patients, and until the late 19th century, there was no better treatment for those patients presenting with syphilis.

Robert Hooke Introduces the Microscope to Study Very Small Organisms

The first demonstration of a world of the very small was presented by Robert Hooke in 1665 with his publication of *Micrographia*.[2] To diarist Samuel Pepys (1633–1703), it was the most remarkable book he had ever read. Hooke was also a gifted artist and his careful drawings of fleas, stem sections of plants, and the

HIERONYMVS   FRACASTORIVS

Also known by his Italian name, Girolomo Fracastoro (1478–1553) is best known for his spore theory of infectious diseases. He received his medical education in Padua, and one of his classmates was Copernicus. Fracastoro believed that linens and clothing carried microscopic spores and caused disease. He named syphilis and successfully used mercury as a treatment. It took more than 300 years before Pasteur proved Fracastoro's theory was correct and renamed as the germ theory of contagious diseases.

honeycomb cellular structure of cork bark described the world magnified by about 30- to 100-fold. It barely included the much smaller world that was about to unfold some 20 years later when the Dutch linen merchant from Delft, Antonie van Leeuwenhoek, began a steady flow of letters and illustrations to the Royal Society of London, where Hooke had also worked. It was Hooke's *Micrographia* that had stimulated Leeuwenhoek to make his own microscopes. But Leeuwenhoek found an easier way to make lenses by creating melted beads from heated threads of glass. He selected those with a minimum of aberration, and some could enlarge the specimens observed about 300- to 500-fold. This was a magnitude better than those of Hooke, and it revealed the world of animalcules, as Leeuwenhoek liked to call them. Some were clearly like insects and showed complex structure. Some were mere geometric rods or spheres, possibly fungi or algae or even very large bacteria that he found in his mouth, in the vases for his flowers, or in other parts of his home and city that he explored. He even looked at his own semen and noted swimming sperm, which he did not distinguish from other animalcules.[3]

Although Pepys expressed enthusiasm, most scientists felt no sudden inspiration to shift to microscopy. Some believed the microscope was a toy well into the end of the 18th century. Some believed it was too difficult to use because of the chromatic and spherical aberration that would not be solved until the early 19th century. Whereas some scientists like Marcello Malpighi and Nehemiah Grew (1641–1712) made excellent use of the microscope to study the tissues or cellular structure of larger organisms, a considerable dedication to microscopy was adopted by amateurs who enjoyed finding new species and sending descriptions of these to natural history clubs and societies. The fields of protozoology and microscopic algal and fungal biology did not flourish until the 19th century, when better microscopes and the advent of staining technology, slide preparation, and other histological techniques emerged in the 1850s and 1860s.[4] Scientists like Ernst Haeckel studied foraminifera and radiolarians and drew exquisite drawings of their calcareous shells, siliceous outer shells, and numerous pseudopodia radiating from them. The major emphasis was a working out of their life cycles and relating their roles in marine life. Many were the source of oxygen for the atmosphere or the primary source of food that led to a food chain for higher organisms. Thomas Henry Huxley's famous essay in 1868, "On a piece of chalk," described in detail the ancient layers of dead *Globigerina* that filled the ocean bottoms and eventually became elevated to form the giant cliffs of Dover and the source of the chalk he held in his hand while giving lectures. It was his touchstone to evolution, and he electrified his audience as they saw the illustrations of these microscopic creatures that in astronomical numbers created much of the English land on which their cities were built.[5]

Pasteur Identifies the Roles of Bacteria and Yeast in Fermentation

By the 1870s, the world of microorganisms was shifting to bacteria through the efforts of Louis Pasteur, who showed that they were living organisms and that they did not arise from spontaneous generation and were, in fact, present in the air wherever humans chose to visit, from the clear air of mountaintops to the air surrounding ships far at sea.[6] Pasteur also showed that it was bacteria that replaced yeast cells in vats and barrels used to brew beer and ferment wine. The bacteria shifted metabolism from yeast cells making alcohol to bacterial cells making acetic acid. In the 1880s, Pasteur extended his findings to bacteria that caused contagious diseases. This allowed physicians to search for microbes as the cause of all infectious diseases. Robert Koch (1843–1910) set up the standards for demonstrating that a specific organism in each case was the causative agent that produced the disease. Koch's postulates and the techniques he used to avoid contamination became the basis of the new field of bacteriology.[7]

Microbial Experiments Begin with Abraham Trembley's Study of Hydra

Experimental studies with microbial organisms began in 1744 when Abraham Trembley (1710–1784) studied hydra.[8] Trembley made his living as a private tutor. He was Swiss and lived in Geneva, and he used his own home as a laboratory. He selected for observation the freshwater invertebrate hydra for its acrobatic behavior as it moved along a surface. He did not know what the organism was and wanted to determine whether it was a plant or animal. He devised techniques to cut hydra into pieces and found which components were essential for regeneration of a complete hydra. He studied a variety of hydras that carried the green algae *Zoochlorella* as endosymbionts, and after cutting shapes into paper he wrapped these around bottles and exposed them to the sun. He showed that removal of the paper after a few days showed all the hydras concentrated in the cutout areas exposed to the sun. His most remarkable experiment was turning a hydra inside out and watching it resume its shape. Trembley did not know when he sent his findings on hydra to the Royal Society that Leeuwenhoek had described hydras in 1702, but Leeuwenhoek did not do the experimental approach that Trembley introduced.

Endosymbiosis Leads to Organelle Evolution

The existence of *Zoochlorella* was also noted in a strain of ciliates, *Paramecium bursaria*. These are protozoa with a slipper shape that have rows of cilia to propel them. The relation of chloroplasts in higher plants, zoochlorellae as symbionts in hydras and in paramecia, and the misnamed blue-green algae (which were

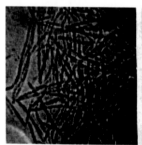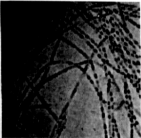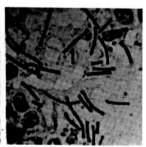

Robert Koch took these photomicrographs of anthrax bacilli in 1876. On the *left*, they are in a filamentous or rod-shaped state and rapidly growing. In the *middle*, they are forming spores. On the *right*, spores are producing the rod-shaped cells. Koch proposed a system (Koch's postulates) that required extracting possible bacteria from blood or other tissue in an infected animal and culturing the cells. The sample of the cultured cells was then injected into a healthy animal. If the animal got sick, Koch would then extract cells from it to show that the same characteristic bacterium was present. Koch obtained pure cultures of the bacilli by growing them on the aqueous humor of the ox's eye. He recorded the multiplication of the bacilli in the aqueous humor culture. When conditions were unfavorable, they produced internal spores. These resisted adverse conditions, such as lack of oxygen. When suitable conditions of life were restored, the spores produced bacilli again. Koch grew the bacilli for several generations in these pure cultures and showed that they could still cause anthrax.

actually a form of bacteria now called cyanobacteria) attracted notice in the 19th century. Andreas Schimper (1856–1901) suggested that chloroplasts may have been derived from blue-green algae.[9] This was more forcefully proposed by Constantin Mereschkowsky (1855–1921) in 1905 to 1909 in a series of publications proposing an endosymbiotic origin of plastids. Mereschkowsky was motivated by years of studying lichens, which are dual organisms composed of fungal cells and algal cells living in a symbiotic relation.[10] In 1920, Ivan Wallin (1883–1969) revived endosymbiosis as a mechanism for evolution and proposed mitochondria as having arisen from bacteria. Other than size and a superficial appearance, there was no evidence for such a role in evolution.[11] A more convincing theory of endosymbiosis as the basis of eukaryotic assemblage from prokaryotic ancestors was developed in the 1960s by Lynn Margulis (1938–2011).[12] She used a biochemical approach to supplement electron microscopy of cell organelles and showed the homologies between mitochondria and bacteria and of chloroplasts with cyanobacteria. Their ribosomes, DNA, and many of their proteins showed this commonality. Originally dismissing her theory as unlikely, most biologists have now accepted the endosymbiotic origin of cell organelles.

Protozoa Reveal They Have a Sex Life

Also of significance in the 19th century was the work of Émile Maupas (1842–1916), a French scholar who made his living as a librarian in Algiers and who used his home as a laboratory for studying protozoa and other microbial species. In 1889, he showed that the ciliate paramecium required some form of sexual activity to prevent dying out.[13] The rejuvenescence involved conjugation in which two paramecia united as if fused and then separated. Herbert Jennings (1868–1947) made major contributions to the biology of protozoa, especially ciliates, during his years at Johns Hopkins University. Jennings worked out a second mode of reproduction in paramecia. He showed that there was a self-fertilization called autogamy, in which the nuclei of paramecia apparently undergo meiosis and form smaller nuclei that fuse and divide, one of the nuclei enlarging as a macronucleus and the other remaining as a micronucleus. It is the micronuclei that engage in the conjugation process when two paramecia swap nuclei.[14] Jennings' student Tracy Sonneborn (1905–1981), at Indiana University, used *Paramecium aurelia* for a series of experiments that revealed their mating types.[15] Other species of paramecium had multiple mating types. Sonneborn also noted the existence of a trait that lead to the death of one of the partners in a conjugal exchange. He called this a killer factor, which was cytoplasmically transmitted. For many years, Sonneborn hoped these would show the existence of what he called plasmagenes. Instead, his students John Preer (1918–2016) and P.K. Chao found they were endosymbiotic small bacteria, highly modified to an adaptive status in the host paramecium.[16] Although plasmagenes have not been proven, something distinctive about the cytoplasm does exist. Sonneborn showed that surgical removal and 180° rotation of a row of cilia led to reverse beating of the cilia with respect to the nonsurgical rows. That persisted after autogamy or conjugation.

Sonneborn's work in the period from 1936 to 1946 was influential in stimulating two fields of biology. It led to a rapid development of microbial genetics using fungi, algae, bacteria, and viruses, and it led to the field of nucleocytoplasmic relations, especially at an experimental level. Most geneticists shared the view that Edmund Wilson proposed in his book *The Cell in Heredity and Development* in 1896. Wilson's view went back to the 1860s with Haeckel's insight into the significance of the nucleus for heredity based on size. Wilson argued that the nucleus with its nuclein and chromosomes and the precision of mitosis and meiosis and the constancy of the chromosomes all argued for hereditary determinants that were located in the nuclei.[17] The cytoplasm was the place where these factors found their expression. Sonneborn depicted this view as a narrow one and doubted that the cytoplasm was merely "the playground of the genes." Sonneborn's colleague at Indiana University, H.J. Muller, had taken Wilson's courses at Columbia and shared Wilson's view. This created both a rivalry and a tension between Muller and Sonneborn but was immensely stimulating for their graduate students who took both their courses and heard their profoundly different views of genetics. Muller won. Plasmagenes failed to show up and κ particles, the causative agent of the killer trait, were modified bacteria. But Sonneborn won something else. Most geneticists were shifting their attention to the work in microbial genetics.

Neurospora Becomes the Means for Working Out Biochemical Pathways

The first effort was performed by George Beadle, who shifted from studying biochemical pathways in fruit flies to biochemical pathways in the bread mold *Neurospora crassa*. In fruit flies, Beadle, working with Boris Ephrussi (1901–1979), used transplants of embryonic rudiments into larvae and showed how eye color in fruit flies involved three components. One was the formation of a brown pigment. The second was the formation of an orange pigment, and the third was a protein on which the pigment was fastened to the eyelets in the compound eye. It was Adolph Butenandt who worked out the chemical components for the brown pigment synthesis, beginning with tryptophan and ending up with kynurenine. He found, but did not name or extend, this gene action to the "one gene–one enzyme" theory that Beadle explored. It was Beadle with biochemist Edward Tatum who realized they could study any simple molecule in living cells—vitamins, amino acids, nitrogenous bases, sugars, and other constituents—by inducing mutations for a specific component in *Neurospora* and identifying through genetic analysis which were precursors of the others.[18] By the early 1940s, biochemical genetics was well established in microorganisms.

Joshua Lederberg Makes Bacteria Model Genetic Organisms

A similar approach was soon performed in bacteria through the work of Joshua Lederberg.[19] He rejected the older view that bacteria divided by a simple fission or separation of the cell into two masses. He showed that the process was more complex and involved the replication and alignment of components, including the single chromosome of the bacterium, before that cell separated into two cells. Also of interest to Lederberg was his discovery that there were mating types to bacteria. One strain donated its chromosome to the recipient. By agitating the conjugating bacteria, he could break that transfer of genes. This allowed him to produce mutations for biochemical traits and then to see how long it took for that mutation to be introduced into a recipient cell. This allowed Lederberg and other bacterial geneticists to map genes by timing gene transfer from donor to host cell. Lederberg was also stimulated by the work going on with bacteriophage. The d'Hérelle bodies of Muller's imagination, in his 1921 paper on "Variations due to change in the individual gene," were becoming test tube creatures subject to the tools of microbiologists. The leading work in phage genetics as it came to be known, was that of Max Delbrück.[20]

The Phage Group Shifts Microbial Genetic to Molecular Biology

Delbrück came to biology as a physicist studying with Niels Bohr (1885–1962). Bohr was stimulated by Muller's induction of mutations by X rays and this introduced physics into biology for Bohr. He believed a physicist's approach to biology would lead to new laws of nature. Delbrück thought he would look for them. First, he went to Berlin where Muller was working on a Guggenheim Fellowship with N.V. Timoféef-Ressovsky. Karl Günther Zimmer, a biophysicist, joined the group at the Kaiser Wilhelm Institute in Dahlem. Muller left for the USSR when Hitler came to power. Delbrück, Timoféef-Ressovsky, and Zimmer tried to study gene properties like size through target theory. Delbrück also hoped to apply quantum theory to the mutation process. All three shared Muller's view that the gene was the basis of life, that genes and viruses could not readily be

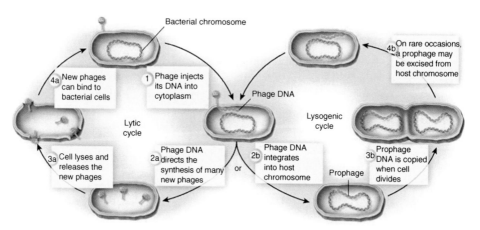

Viruses can infect bacterial cells in various ways. On the *left*, a bacteriophage (green) is on the cell wall of a bacterium. The bacterial DNA is red. The DNA is injected into the cell. If it multiplies the viruses (*left side*), the bacterium will disintegrate (lysis), releasing dozens or hundreds of virus particles. Some strains of bacteriophage are lysogenic (*right side*). The DNA of the virus enters and integrates into the bacterial DNA (a process called lysogeny). When the bacterial chromosome is synthesized and the cell divides, each cell gets the integrated virus. The integrated virus is referred to as being a prophage. A change in the environment may lead to the release of the prophage in the cell and it switching to the lytic phase on the *left*, killing the cell and releasing a shower of new viruses.

distinguished, and that the property that made genes unique was their capacity to replicate their variations. They argued these positions in the 1930s at gene conferences held at Spa in Belgium. They wrote a paper in 1935 offering the results of their target theory research and Delbrück's quantum-based gene mutation, which famously became known as the "Three-Man Paper."[21]

In 1936, Delbrück, disgusted with Nazi policies, left Germany and came to the United States. In 1937, he went to Caltech and teamed up with Emory L. Ellis (1906–2003), who was working with bacteriophage, to work out the viral life cycle. They showed that the host cell gets infected and no viruses are released until a lysis occurs, resulting in a dramatic increase in progeny. They broke open host cells that were infected and showed that there was a latent stage with no mature virus formation during that latency. The numbers rose and finally a massive release occurred with the lysis of the host cell wall. Delbrück then joined Salvador Luria, and they soon identified a number of morphological mutations involving plaque size and shape on the lawn of *Escherichia coli* host cells of the Petri dishes used to study the viruses. Delbrück and Luria mated concentrations of viruses that would permit multiple infections of cells, some of which would receive two different phage mutations, like turbid and smaller rapid lysis plaques. They obtained recombinant viruses that were normal for both traits and recombinants that had both rapid lysis and turbidity as traits expressed in a

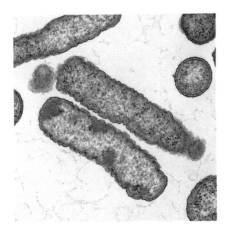

This transmission electron micrograph of bacteria shows rod-shaped (called bacilli) *Elizabethkingia anophelis* bacteria. Other bacterial forms are spherical-shaped (called cocci) and corkscrew-shaped (called spirilla). Bacteria sometimes have an elongated tail called a flagellum that permits motion.

plaque. By counting the number of these recombinant events, they could establish a map of viral gene mutations.[22] These bacteriophage results from 1937 to the mid-1940s led Lederberg to explore a possible genetics of *E. coli* bacteria, and he found this through the techniques he developed for isolating mutations and for finding sexual strains in bacteria.

The Field of Microbiology Arose by Incremental Changes

It is interesting to reflect on how ideas are transmitted and how they shift thinking in new fields and transform old fields. Haeckel's theory of the nucleus as the site of hereditary factors in 1866 shifts into Wilson's school about 1900–1906, leading to a chromosome theory of heredity. Muller's fantasy of genes studied in a test tube and genes as the basis of life in the 1920s shifts as a new field of virus genetics becomes a reality through absorption by Delbrück in the 1930s. Delbruck's contribution as a physicist in biology gets shifted in 1945 into the theory of *What Is Life?*, suggesting an aperiodic crystal and a codescript for the gene by physicist Erwin Schrödinger, who absorbs these ideas through the work of Timoféef-Ressovsky, Delbrück, and Zimmer.[23] It shifts again from Schrödinger into the thinking of Watson and Crick in 1953 in the working out of the structure of DNA and what it means—the nucleotide code as genes and as information.

Science is a vastly interconnected web with ideas resonating across contemporaries and across time, sometimes a century or more apart, and rejuvenated by new techniques, new organisms, and fresh enthusiasm. The revolutions are not paradigm shifts. They are incremental additions to microbial genetics that have their roots well into the past. They could not have happened without the

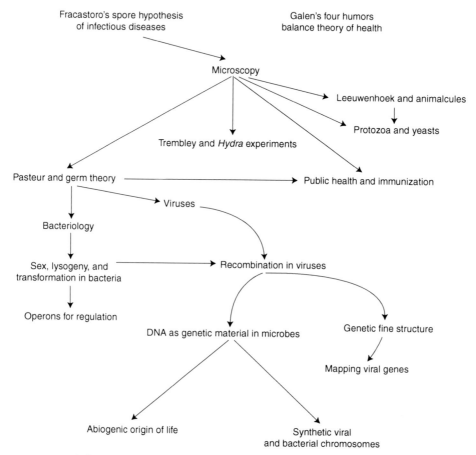

Fracastoro believed microscopic spores caused infectious diseases. He rejected Galen's theory of systemic disease using four humors in a balanced state. Fracastoro did not know if the spores were alive and no microscopes existed to see them. Leeuwenhoek was the first to see microscopic living organisms, such as rotifers, ciliates, and amoeboid cells. He called these animalcules (and included sperm among the animalcules). Abraham Trembley cut hydras and showed they regenerated their missing parts. He also recognized strains of hydra with endosymbiotic algal cells. Pasteur used the microscope to identify the cause of wine and beer spoilage. He distinguished rod-shaped vinegar-making bacteria from round, larger yeast cells involved in fermentation of sugars to alcohol. Pasteur applied his knowledge of bacteria as germs to foster public health through immunization. Koch worked out the procedures for a field of bacteriology. Viruses were detected in 1890 for tobacco mosaic virus and for bacteriophage in 1914. In the 1940s, bacterial transformation was found. In the 1950s, bacterial sex, lysogeny, and genetic recombination were found. Also in the 1940s, viral recombination was worked out along with gene maps of viral chromosomes. Regulation of genes was first studied with the operon model. Theories of the origin of life shifted from vague assemblies to the first formation of RNA after a category of RNA molecules was shown to have enzymatic function. Recent studies show artificial life is possible and both viral DNA and bacterial DNA have been synthesized and shown to be functional.

incremental introduction of new techniques, new organisms, new tools, and the inspired insights that tweaked the past insights into new levels of implications and findings.

References and Notes

1. Fracastoro G. 1546. *De contagione et contagiosis morbis et eorum curatione. Libri III.* [*On contagion, contagious diseases, and their cure.*] See also Pachter HM. 1951. *Magic into science: The story of Paracelsus.* Schumann, New York.

2. Hooke R. 1665. *Micrographia.* Royal Society, London.

3. Dobell C. 1960. *Antony van Leeuwenhoek and his "little animalcules."* Dover Reprint, New York.

4. Conn HJ. 1933. *The history of staining.* Biological Stain Commission, Geneva, New York. Also see Haeckel E. 1899–1904. *Kunstformen der Natur.* [*Art forms in nature.*] Verlag des Bibliographisches Institut, Leipzig and Vienna.

5. Huxley TH. 1868. On a piece of chalk. In *Collected essays.* 1908. MacMillan, London.

6. Pasteur L. 1866. *Études sur le vin, causes qui les provoquent, procédés nouveaux pour le conserver et pour le viellir.* Imperiale, Paris.

7. Koch R. 1880. *Investigations into the etiology of traumatic infectious diseases.* The New Sydenham Society, London.

8. Baker JR. 1952. *Abraham Trembley of Geneva.* Arnold, London.

9. Schimper A. 1885. Untersuchungen über die Chlorophyllkörper und die ihnen homologen Gebilde. *Jahrb Wiss Bot* **16**: 1–247.

10. Mereschkowsky C. 1905. On the nature and the origin of chromatophores (plastids) in the plant kingdom. In Martin WJ, Kowalik KV. 1999. Annotated English translation of Mereschkowky's 1905 paper Über die Natur und Ursprung der chromatophoren im Pflanzenreiche. *Eur J Physiol* **34**: 287–295.

11. Wallin I. 1923. The mitochondria problem. *Am Natur* **57**: 255–261.

12. Sagan LM. 1967. On the origin of mitosing cells. *J Theor Biol* **14**: 255–261.

13. Maupas E. 1883. Contribution a l'étude morphologique et anatomique des infusoires ciliés. *Arch Zool Exp* **1**: 427–452.

14. Jennings HS, Lashley KS. 1913. Biparental inheritance and the question of sexuality in paramecium. *J Exp Zool* **14**: 393–466.

15. Sonneborn TM. 1937. Sex, sex inheritance, and sex determination in *Paramecium aurelia. Proc Natl Acad Sci* **23**: 471–502.

16. Preer JR. 1950. Microscopically visible bodies in the cytoplasm of the "killer" strain of *Paramecium aurelia. Genetics* **35**: 348–362.

17. Wilson EB. 1896. *The cell in inheritance and development.* Macmillan, New York.

18. Beadle GW, Tatum EL. 1941. Genetic control of biochemical reactions in *Neurospora. Proc Natl Acad Sci* **27**: 499–506.

19. Lederberg J, Tatum EL. 1946. Gene recombination in *E. coli. Nature* **58**: 588.

20. Ellis EL, Delbrück M. 1939. The growth of bacteriophage. *J Gen Phys* **22**: 365–384.

21. Timoféeff-Ressovsky NV, Zimmer K, Delbrück M. 1935. Über die Natur der Genmutation und der Genstruktur. *Nachr Ges Wiss Göttingen: Math Phys Klasse, Fachgruppe VI, Biologie* Bd 1 Nr. 13: 189–245.

22. Luria S, Delbrück M. 1943. Mutations of bacteria from virus sensitivity to virus resistance. *Genetics* **28**: 491–511.

23. Schrödinger E. *What is life?* Macmillan, New York.

CHAPTER 10

Embryology: From Philosophic Forms to Epigenetic Organogenesis

Epigenesis versus preformation. Sectioning embryos, descriptive embryology and 3D reconstructions. Roles of growth, differentiation, and morphogenetic movements, *Entwicklungsmechanik* and the growth of experimental embryology. Holism versus reductionism. Twinning, nuclear cloning, chimeras and embryonic gene expression. Homeotic genes and body plan and symmetry development.

Until the late 18th century, women who were pregnant first noticed a missed menstrual period and then morning sickness, both signs coming in the first third of the approximately nine months of gestation. Enlargement of the breasts, followed by "quickening," the first evidence of motion of the new life in the womb, characterized the second three months. The rapid growth of the abdomen and contractions immediately before the birth process itself were part of the last third of that pregnancy. Theologians argued about the moment ensoulment occurred. For some it was at birth. For some it was at quickening. Only after the cell theory and the working out of fertilization in the 1860s–1870s did theologians begin to believe conception was the moment of ensoulment.[1] This became even more controversial with the analysis of twinning in the early 20th century, with monozygotic twins arising anywhere from a two-cell stage to an implanted blastocyst with embryonic tissue layers present. Severe birth defects, like two-headed babies or chimeric individuals resulting from the fusion of two fertilized eggs, challenged theological classification. Not only was the introduction of descriptive embryology a problem, so was the idea of conception. Did it mean fertilization of an egg (unknown before the 1870s)? Did it mean the time of implantation of a blastocyst into the uterine lining (unknown until the late 19th century)?

The Problem of Personhood before Microscopy

Embryology also presented problems for philosophers as well as theologians who tried to determine when legal status as a person should be assigned. A lot of those debates are recent. Is a zygote a person? Is a blastocyst? Does it have a specific time in organogenesis, perhaps when the brain reaches a certain stage of anatomical development? Or is personhood a process rather than an event, reaching legal status at birth when virtually everyone agrees that a live birth is a person and most people consider a dead baby at delivery a stillborn who has had no legal status as a person until recently. Some U.S. states are now issuing birth certificates and death certificates for stillbirths.[1] Those human issues did not apply to the status of chickens, frogs, or pigs that underwent embryological development. Personhood and ensoulment are human concerns.

In the nonhuman animal world, it is a question of science that decides when events occur and what constitutes process rather than essential status. Aristotle was the first to study embryology using chicken eggs, and he followed the daily changes over the 21-day gestational cycle. For Aristotle, the egg was an external uterus and the shell could be broken to reveal the inmost parts of the developmental process. He saw first a clot of blood as evidence of a future chicken. Form came incrementally as he looked at the tiny structures, some recognizable as nascent organs like hearts and others that were absent from the adult (like the extra-embryonic membranes) and that were too vague to be classified. Aristotle assumed the male provided the form; the female provided the shapeless mass that was the unfertilized egg. Those unfertilized eggs were "wind eggs" in the minds of farmers collecting eggs for their family or for sale in the markets. He concluded that embryonic development was epigenetic, his term to describe a gradual emergence of form that was not predictable or present at the initial stage of embryonic development.[2]

Aristotle's view prevailed for about 2000 years. It was challenged in the late 17th century as microscopes began to appear and as attention was paid to the detailed anatomy of microscopic creatures. Jan Swammerdam (1637–1680) in the Netherlands was one of the first to revive the debate over the Aristotle–Harvey view of epigenesis and a growing belief that this was inadequate. Swammerdam studied insects and worked out the metamorphic stages of flies and butterflies. He showed that there was a deposited egg, a larval stage that was wormlike, a pupal stage in which epigenetic development occurred, and finally the emergence of the adult whose form could be seen emerging in the pupa. It had to be epigenetic because the egg, larva, pupa, and adult were not four separate animal species but all phases of one insect followed for its life cycle. Swammerdam's major biological work was reported in 1669, but he left the field in 1675 to devote the rest of his life to religious mystic contemplation.[3]

Preformation and Epigenesis Contended in the 17th Century

Challenging the epigenetic view was the work of Nicolas Malebranche (1638–1715). He was a French priest who admired the work of René Descartes

(1596–1650) and St. Augustine. He broke with the scholastic tradition and proposed a theory that ideas were preexistent in God's mind and this extended to nature itself. Scientists studying nature were actually studying archetypes of something that was everlasting and traceable back to God's mind. He came up with the idea that all living things flowed from this divine origin and were encapsulated in the present to be released into the future. This was the origin of the Russian doll model of preformation; each layer of an egg had another egg layer behind it in an immense regression to the Second Coming of Christ and the end of life and procreation as we know it. Malebranche offered it in his *Treatise on Nature* in 1680. This set up the preformationist–epigenesis debate on embryonic development.[4]

Preformationists split into two camps. Ovists argued, like Malebranche, that it was the egg that had this encapsulated set of future generations. Spermists argued that such encapsulated sets resided in the sperm. Both agreed that the process of development was a mechanical one of enlargement. They also differed on the philosophic implications of their position. Some were deists, and they believed God set the universe into action like a gigantic clockwork that needed no further day-to-day action by God for the universe to work by divine law. These scientists believed the world was rational and the material world was distinct from the spiritual world. Others believed that the preformation of life supported a vitalistic and not a materialistic interpretation of life. The structure of life could be described to an extent by anatomy and other approaches, but there were limits on how far it could be resolved by reductionism.

Each side had a rough time reconciling scientific interpretation and prevailing theology. One issue for the preformationists was the massive quantity of sperm present in an ejaculate. If the sperm had the preformed individual of all future generations, why were virtually all of them wasted in an ejaculate? Was this not contradictory to a God who cared for all his creations? What was the role of the sperm for ovists? How did any theory of preformation account for the apparent equal contribution of both partners to such traits as skin color in interracial couples? How did preformationists explain Jan Swammerdam's finding of metamorphic stages in insects? Enlargement was insufficient for such markedly different stages of development. How did those favoring epigenesis explain the formation of shape and position and organization of the body components from apparently formless material preceding their appearance? Was not epigenesis the real vitalism, whereas preformation was consistent with scientific reason and practice?

The evidence for ovists came from the work of Charles Bonnet (1720–1793), a Swiss physician.[5] His studies of parthenogenesis in aphids showed that males were not involved in aphid production. By default, all aphids had to come from virgin females. Bonnet noticed that there were smaller aphids developing in the abdomens of adult aphids. They looked just like the mother but much smaller. This was evidence for the ovist's position. Bonnet also proposed a theory of God's creation in which he laid out a "scale of being" from the smallest and lowest forms of life to human life as the present apex of God's handiwork. At the time of the Second Coming, there would be a higher level (perhaps angel-like) awaiting human development in this scale of being. It was not quite an evolutionary argument, but it fed the rational desire for meaning. Most scientists of that age believed in a rational God, and the scale of being satisfied that mental expectation. A role for science was working out that scale of being. It would be like reading the mind of God. This fit an emerging belief among most scientists of that era that reading the mind of God could be performed through the "Bible of nature" to supplement the scriptural word of God in the Old and New Testaments.

Spermists were ridiculed by ovists and epigenesists. The two most publicized portraits of homunculi in sperm are probably hoaxes. Delanpatius is a pseudonym of one of the hoaxers. He drew the tiny figures in several isolated sperm as shadow figures. In 1694, Nicolaas Hartsoeker (1656–1725) drew an elaborate

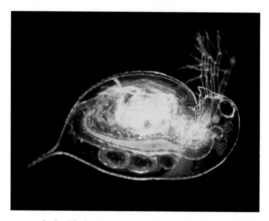

Parthenogenesis means virgin birth. Some organisms reproduce parthenogenetically (rotifers, aphids, monitor lizards), and mostly in these cases the females produce daughters. In some species, there is a sexual cycle that can be turned on or off by environmental conditions. In parthenogenetic strains lacking a sexual cycle, gene flow is often from partially digested DNA of dead or cannibalized members of that species or genes from other organisms they digest. This photo of *Daphnia pulex* shows three immature daphnia in her brood pouch. Most crustaceans lay fertilized eggs and development is outside the female's body. Bonnet observed a similar presence of aphids in a parthenogenetic mother and believed he had found evidence that life is preformed and not generated by epigenetic events leading to greater complexity.

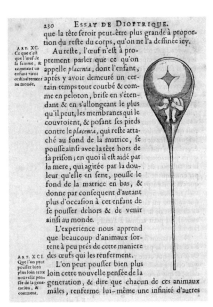

Nicolaas Hartsoeker was a Dutch lens maker and student of Leeuwenhoek. He codiscovered sperm independently of Leeuwenhoek. His principal work was making instruments, especially microscopes and telescopes. He rejected Hooke's belief that if one made a telescope big enough one could see people walking on the moon's surface. In his account, Hartsoeker never claimed to see a little man in the sperm. A similar satirical spoof of the preformationists was made by Delanpatius, a pseudonym for François de la Plantade, who published illustrations of fake sperm with little people in them.

homunculus in a sperm head with detail that no microscope of that day (including his) could have revealed. Hartsoeker made it clear that he did not actually observe the *"petit l'enfant."* Rather, he suggested that this is what could be seen if there was a way to peer inside the sperm.[6] At best, he drew a detailed figure suggesting what to look for if such homunculi existed. Because he also disproved Hooke's belief that a refracting telescope might reveal humans walking on the moon, it is more likely that Hartsoeker drew this illustration to show the absurdity of the spermist claims.[7] He was a physicist who made instruments and invented the screw barrel for adjusting the focus of a microscope's lens. The trouble with hoaxes is that they make sense at the time but as generations pass by, the work is taken as evidence of gullibility or self-deception by the hoaxer. When Friedrich August Kekulé proposed a ring-shaped benzene molecule with carbon atoms in a hexagon, one chemist sent a letter to the journal claiming he had a similar model of six chlorine atoms in a hexagonal shape. He signed it, S.C.H. Windler. Unfortunately for the skeptical "swindler," the reality of the benzene ring prevailed. Fortunately for the field of embryology, spermists and ovists disappeared.

Caspar Wolff Demonstrates Epigenesis with Improved Microscopy

Classical embryology and definitive epigenesis were established initially by the work of Caspar Friedrich Wolff (1733–1794), a German biologist who got his

M.D. at Halle.[8] He studied plant and animal development and shared with Harvey a belief in vitalism as the basis of the emergence of form in the embryo. In 1768, he made a detailed study of the chick embryo and noted what would later be called the endoderm. He described a leaflike layer of tissue that formed a gut tube, with a gradual emergence of stomach, pancreas, liver, and gallbladder, as well as the various regions of the intestines. He also followed the development of the kidney and described the mesonephros and the adjacent duct named for him, the Wolffian ducts. His work stimulated Heinz Christian Pander (1794–1865), who was born in Riga and spent his life in Germany, the Baltic States, and Russia carrying out his work. The first phase of his scientific life was in embryology. He identified the three germ layers (mesoderm named by Huxley in 1871, ectoderm named by E. Ray Lankester [1847–1929] in 1873, and endoderm also named by Lankester in 1873) and followed them in the chicken.[9] He noted that blood vessels formed from the middle layer.

The third founder of classical embryology was Karl Ernst von Baer (1792–1876) who was Estonian, obtaining his M.D. in Dorpat.[10] His career put him in Estonia, Germany, and Russia. In 1826 he worked on embryology through the influence of Ignaz Döllinger (1770–1841). He studied chicken, frog, and mammalian embryos, confirming the epigenetic basis of development in animals. He confirmed the three layers that Pander had found and pursued these with detail. In 1826, he discovered the mammalian egg in the Graafian follicle and he described the fetal membranes and their relation to the extraembryonic membranes. He noted the formation of the blastula in frog development. He also described the notochord as a feature of vertebrate development of the spinal column. He summed up all the microscopy evidence and wrote what can be considered the first textbook of embryology in 1828. von Baer drew an important conclusion from his studies of embryos. He claimed that the embryo of a higher animal is never like the adult of a lower animal. It is only like its embryo. At odds with this interpretation was Ernst Haeckel's claim in 1866 that "ontogeny recapitulates phylogeny." Both are reconciled by an evolutionary argument that the basic body plan of earlier life became the developmental starting point for more advanced modification.

Wilhelm Roux Introduces *Entwicklungsmechanik*

The shift from descriptive embryology to experimental embryology began in the last quarter of the 19th century. Earlier efforts producing regeneration in the hydra experiments of Abraham Trembley did not lead to an active school of experimentalists. The major difficulties for embryos were their small size and the need for effective microscopy to describe and manipulate events at the cellular level for the earliest stages of embryogenesis. The first to do this was Wilhelm Roux (1850–1924). He was born in Jena and studied with Ernst Haeckel there for his M.D. He also studied with Rudolph Virchow in Berlin. His dissertation for his M.D. was on the embryonic development of blood vessels. It was Roux in 1885 who introduced what he called "*Entwicklungsmechanik*" (or developmental mechanics) implying physiological/experimental approaches to embryological studies or by disrupting normal developmental processes and studying their effects. It is today called experimental embryology.[11] He used frog embryos because the eggs they laid were large and allowed the fertilized eggs to divide to the two- or four-cell stage. He then destroyed half of the cells in each experimental run, leaving the other remaining blastomere or blastomeres of an embryo intact. In each case, he got a half-embryo developing. To Roux, this

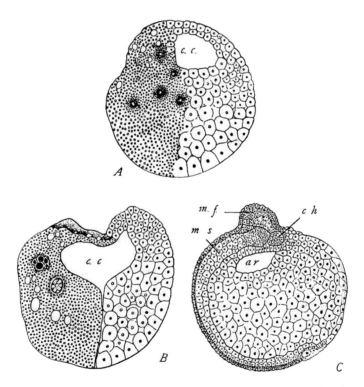

Wilhelm Roux used a hot needle to cauterize one of two blastomeres of a divided frog zygote (A). In B, the surviving half-embryo is differentiating its tissues. In C, the half-embryo partially restores a body plan, but it will lead to a deformed tadpole. Roux believed the first division of the zygote sorted out different structures to each blastomere. That interpretation turned out to be incorrect. If the two blastomeres are separated so that each is alive and not damaged, they will form normal-looking tadpole twins.

suggested that the first two cells sorted out different chemical components and that differentiation had begun with this first cell division. He called this mosaic development, and in the view of his critics he had revived preformation. In 1888, he also isolated a segment of the chick embryo (the medullary plane region of the embryo just before formation of its three layers) and kept it alive for several days in a saline solution. It was the first attempt at tissue culture.

Blastomeres, Totipotency, and Mosaic Development

Opposing Roux's interpretation of mosaic development was Hans Driesch (1867–1941). He studied with August Weismann at Freiburg and with Haeckel and Otto Hertwig at Jena. In 1895, he used sea urchin eggs to study their embryonic development and test Roux's findings. He found to his astonishment that if he separated the two cells of the first division after fertilization, each blastomere formed a separate normal sea urchin.[12] He had produced artificial twinning by cloning. He then tried shuffling the eight cells of an eight-celled stage morula and found that however they were randomly shifted about, the reconstituted morula went on to form a perfectly normal adult. This suggested to Driesch that the blastomeres were "totipotent," his term for retention of the capacity to produce an entire individual from its part. It was not strictly like regeneration because the entire organism seemed to be potentially present in each of its cells. This forced Driesch to adopt a vitalistic interpretation of life. He called this living potential an "*enteleche*" (or entelechy in English), resurrecting Aristotle's views. It also led him, after 1909, to shift his career from biology to philosophy, and he taught and wrote about philosophy from then on until Hitler came to power. Driesch was a pacifist and he also objected to Nazi policies. Although Driesch had no Jewish ancestry, he was the first non-Jewish scientist forced out of university teaching by the Nazi purge of non-Aryan or anti-Nazi personnel at the university level. One of the ironies of human capacity for friendship and its independence from rational understanding is Driesch's lifelong friendship with Thomas Hunt Morgan, whom he had met in the Naples Station while Morgan was doing the "grand tour" before settling into Bryn Mawr for his early career. Morgan had revulsion to vitalism, holism, and other supernatural interpretations of the life sciences; he was committed to reductionism, materialism, and mechanism as the legitimate and most effective outlooks for experimental science.

Hans Spemann (1869–1941) tried a different experimental approach. He used a fetal hair to isolate half the cytoplasm from the nucleus of a zygote. The nucleus divided by mitosis to produce a 16-cell stage. He then released the fetal hair and allowed one of the nuclei to enter the isolated cytoplasmic

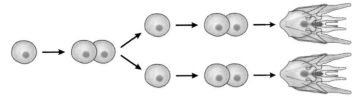

Hans Driesch in 1891 used sea urchin eggs to do experiments stimulated by Roux. In Driesch's approach, he produced two-celled embryos and shook the flask violently to separate most of the embryos. He then took single cells of these and allowed them to grow. To his surprise, they formed perfectly normal larvae and matured into adult sea urchins. This suggested to Driesch that the body plan is present in each cell. He believed there was a unique life essence present that he called an *enteleche*. He called the study of these life essences vitalism. Surprisingly, he and T.H. Morgan got to know each other when Morgan visited Europe. They remained close friends despite Morgan's utter rejection of vitalism and his embrace of reductionist materialism as his guide for interpreting life. Driesch's demonstration of artificial twinning by blastomere separation eventually led to stem cell research and nuclear transplantation from frogs to sheep.

mass. This now nucleated blastomere was detached and produced a normal embryo and tadpole, proving that early blastomeres were totipotent.[13]

The evidence that changes eventually take place in the nuclei of cells during development came from studies of Robert Briggs (1911–1983) and Thomas Joseph King (1921–2000) while they were at Fox Chase Institute in Philadelphia in 1952. They showed that nuclei of blastulas were totipotent but after gastrulation they became incapable of generating a normal embryo or adult offspring. John Gurdon (b. 1933) extended their work. He was the first to clone an animal from a single adult cell. He cloned a frog using the intact nucleus from a somatic cell of *Xenopus*. This work eventually led to the first cloned mammal, Dolly the sheep (1996–2003), by somatic cell nuclear transfer in 1996, by Ian Wilmut (b. 1944) and his team in Edinburgh, Scotland.

The Union of Embryology and Genetics

The last phase of experimental embryology was its eventual union with genetics, biochemistry, and molecular biology. This did not begin until the mid-1930s. When Morgan published his book *Embryology and Genetics* in 1934, Richard Goldschmidt (1878–1958), who gave it a negative review, said the book lived up to its title. Half of the book, he claimed, was about embryology and half the book was about genetics, but nowhere in the book were the two fields connected. That changed when George Beadle working with Boris Ephrussi in Morgan's department at Caltech worked out the sequence of events associated with eye color development in fruit flies. They used transplants of eye rudiments found in the late larval stage and placed these in other larvae of a different genotype.[14] Thus, a vermillion-eyed rudiment in a normal red-eyed larva produced a normal red eye. It was converted by some diffusible substance in the host larva. But if a white-eyed rudiment was placed in a red-eyed larva, the transplanted eye was white-eyed and it had retained its autonomy. When trying these transplants with different members that lacked the brown pigment, some gave red-eyed development and some retained their mutant bright orange-colored eye. Beadle and Ephrussi recognized that their embryological approach had shown biochemical pathways exist and gene action could be described as a sequence of enzymatic events that started with a simple product and led to an increasingly more complex molecule. Very quickly, Beadle recognized that the genetic implications of this work outweighed the embryological ones, and he switched to using the red-orange bread mold *Neurospora crassa* to work out biochemical genetics and open up microbial genetics for exploring gene action. The fusion of embryology and genetics did not take place for another generation. It came from the work of another Caltech student working in the fly lab now headed by Alfred Sturtevant. Edward B. Lewis grew up in Wilkes-Barre, Pennsylvania, and attended the University of Minnesota, where he worked with Clarence P. Oliver who encouraged his fruit fly interests.

Oliver recommended Lewis for a fellowship to Caltech, where Lewis got his Ph.D. with Sturtevant. He served in the Air Force and returned after the war to Caltech. There he began a long commitment to the field of pseudoallelism and he shifted from the star-asteroid and white-eyed series to bithorax. The bithorax series of mutations were homeotic. They involved deformities of the wing, abdomen, thorax, and halteres. Lewis showed over the years that these mutations formed a linear series with a likely effect on the repeated boxes that are characteristic of arthropod development.[15] His work in 1978 showed that the genetic sequence found by DNA analysis and the map sequence that he worked out matched and that homeotic mutants worked by diffusible substances affecting the anterior or posterior ends of these boxes leading to thorax, haltere, wing, and abdomen differentiation. Lewis could generate four-winged flies, eight-legged flies, or wingless flies. His work united genetics, embryology, and evolution.

Developmental Biology Shifts to Molecular Biology through Regulatory Genes

François Jacob (1920–2013) and Jacques Monod (1910–1976) showed in bacteria that adaptive enzyme formation was associated with a special category of genes that used an operator, its protein product, and the recipient gene or genes it regulated. They referred to the collective genes involved as an operon.[16] The operon model explained how cells could shift from digesting glucose to digesting lactose. It also explained how metabolism could be regulated in the cell. By inference, Jacob and Monod anticipated similar regulatory genes must exist for embryonic development. At the time (1957–1960) when operons were being studied for metabolism, there were no suitable models of embryonic regulatory genes. That changed a generation later when E.B. Lewis applied the operon model to fruit flies using a homeotic mutant, bithorax.

The molecularization of homeotic mutations was worked out independently in 1983 in Walter Gehring's (1939–2014) laboratory in Basel and by Matthew P. Scott and Amy J. Wiener in Thom Kaufman's laboratory at Indiana University.[17] They identified the homeobox as a region of 180 base pairs that produced a protein that binds to DNA of genes involved in the anteroposterior development of the embryo. In insects, this involves the segments of the head, thorax, and abdominal regions as well as their associated structures (legs, wings, halteres). The first vertebrate homeobox was isolated by Edward de Robertis' (b. 1947) laboratory using the clawed frog *Xenopus*.[18] The field of evo-devo in evolutionary biology is an outcome of these findings of developmental genetics and its relation to the evolution of wing and leg formation in insects. It immediately set up the possibility about how segments could produce new structures, convert preexisting structures, or repress the formation of repeated structures, all of which play a role in the origin of insects from earlier arthropods.[19,20]

Developmental Biology Follows an Incremental Model

The passage of development from its effects in human pregnancy to the first attempts to look at embryos with the unaided eye by Aristotle led to a theory of epigenetic development in which the mystery of emergent form encountered a permanent barrier associated with vitalism. That shifted as preformation contended with epigenesis and embraced first mechanism and then vitalism in the 17th and 18th centuries. By the 19th century, with the refinement of microscopy, the epigenetic basis of organogenesis was worked out. Embryology acquired an experimental approach initially from the work of Trembley for the study of regeneration of parts from a mutilated organism, but at the cellular level experimental embryology got its start in the 1890s with the works of Roux and Driesch using *Entwicklungsmechanik*. Embryology fused first with genetics, then biochemistry, and then evolution. In its most recent transformation as a science, it has become molecular with the discovery of homeoboxes and their gene sequences and their roles in anteroposterior development of all known animals. It is difficult from this historical summary to infer major paradigm shifts.

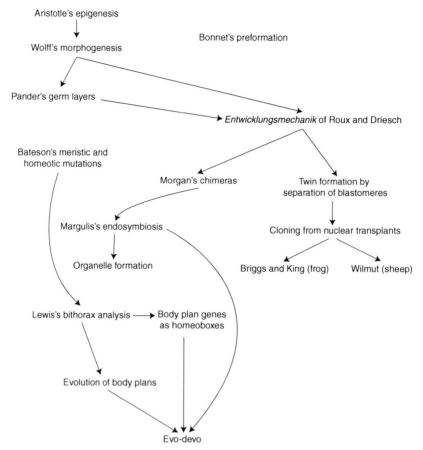

Aristotle opened chicken eggs each day from one-day-old eggs to eggs that began to hatch. He showed that embryonic form was not a simple miniature of the chick. His view is called epigenesis. That view prevailed until the Age of Enlightenment when Bonnet proposed a simple mechanism to account for development—growth. From microscopic size to birth, the individual was shaped like a person or the organism characteristic of the egg source. Preformation was usually maternal (ovists), but some were spermists. Wolff showed that Aristotle's observations were correct, and he followed the formation of each major organ using a microscope. Pander introduced the idea of embryonic germ layers (ectoderm, mesoderm, and endoderm). Bateson claimed there were mutations that he called meristic (extra parts) or homeotic (displaced location). German scientists introduced *Entwicklungsmechanik*, which allowed scientists to do experimental embryology. This led to artificial twin production by blastomere separation and later the use of nuclear transfer from donor cell to enucleated egg. This was first performed by Briggs and King with the frog *Rana pipiens* and later by Ian Wilmut to produce Dolly the sheep from the nucleus of a breast cell. Lewis studied the bithorax gene in fruit flies and worked out the relation of the genes to wing, halteres, and abdominal segments. His work led to the discovery of homeobox (Hox) genes. This also led to the field of evo-devo and united evolution and embryology. Cloning also led to the study of stem cells and their use in organ replacement or repair.

More often in this field, the view of vitalism and reductionism as philosophic (or religious) attitudes are imposed on the biology, which oscillates between the two outlooks. If these religious and philosophic aspects are omitted, the scientific contributions are supportive of incremental changes. With the introduction of genetics, biochemistry, and molecular biology into developmental biology, the location of vitalistic views is becoming more remote and reductionist explanations are presently in the ascendant for how we progress from a fertilized egg to an adult being.

References and Notes

1. http://www.nytimes.com/2007/05/22/us/22stillbirth.html.

2. Aristotle. ca. 350 BCE. *De anima.* Trans. Thompson DW. *The history of animals.* Book VII. Available online at http://classics.mit.edu/Aristotle/history_anim.7.vii.html.

3. Swammerdam J. 1669. *Historia insectorum generalis.* Latin trans. Henninius HC. 1733 reissue. Abkoude, Leyden, Netherlands.

4. Detlefson K. 2004. Supernaturalism, occasionalism, and preformation in Malebranche. *Persp Sci* **11**: 442–487.

5. Bonnet C. 1782. *Contemplation de la Nature.* [*The contemplation of nature.*] J.G. Virchaux & compagnie, Hambourg.

6. https://embryo.asu.edu/pages/nicolaas-hartsoeker-1656-1725/.

7. Hill KA. 1985. Hartsoeker's homunculus: A corrective note. *J Hist Behav Sci* **21**: 178–179.

8. Wolff C. 1774. *Theoria generationis.* Hendall, Halle, Germany.

9. Pander HC. 1817. *Beitrage zür Entwicklungsgeschicte des Hünchens im Eye.* Brömer, Würzburg, Germany.

10. Von Baer KE. 1828. *Entwicklungsgeschicte der Theire. Beobachtung und Reflexion.* Bornträger, Königsberg, Germany.

11. Roux W. 1888. On the artificial production of half embryos by the destruction of one of the first two blastomeres. In *Foundations of experimental embryology* (ed. Willier BH, Oppenheimer J), 1975, p. 36. Hafner Press, New York.

12. Driesch H. 1908. *The science and philosophy of the organism.* pp. 38–42. A. and C. Black Publishers, Ltd, London.

13. Spemann H. 1938. *Embryonic development and induction.* Yale University Press, New Haven, CT.

14. Beadle GW, Ephrussi B. 1935. Transplantation in *Drosophila. Proc Natl Acad Sci* **21**: 642–646.

15. Lewis EB. 1978. A gene complex controlling segmentation in *Drosophila. Nature* **276**: 565–570.

16. Jacob F, Monod J. 1961. Genetic regulatory mechanism in the synthesis of proteins. *J Mol Biol* **3**: 318–356.

17. McGinnis W, Levine MS, Hafen E, Kuroiwa A, Gehring WJ. 1984. A conserved sequence in homeotic genes of the *Drosophila* Antennapedia and bithorax complexes. *Nature* **308:** 428–433.

18. Scott MP, Weiner AJ. 1984. Structural relationships among genes that control development: Sequence homology between the Antennapedia, Ultrabithorax, and fushi tarazu loci of *Drosophila*. *Proc Natl Acad Sci* **81:** 4115–4119.

19. De Robertis EM, Sasai Y. 1996. A common plan for dorso-ventral patterning in bilateria. *Nature* **380:** 37–40.

20. Carroll S. 2005. *Endless forms most beautiful: The new science of evo devo and the making of the animal kingdom.* W.W. Norton, New York.

CHAPTER 11

Cell Organelles: From Cell Theory to Cell Biology

The cell wall, protoplasm, nucleus, cytoplasm, mitochondria, endoplasmic reticulum, Golgi, lysosomes, nuclear envelope, cell membrane, plastids, endosymbiosis, chromosomes, ciliary bodies, macronucleus versus micronuclei, centrosomes, spindles. Tools for organelle functions: centrifuge, radioisotope labeling, chromatography, microscopic surgery.

We have seen how the cell theory emerged from Hooke's empty boxes in cork to the protoplasmic models of the mid-19th century and to the chromosome distributions leading to mitosis and meiosis in the last two decades of the 19th century. The 20th century was largely an effort to identify the organelles of the cells and the structures and functions of those organelles. It led to the field of cell biology, which is more experimental than it is descriptive. The term "organelle" was first suggested by Karl Möbius (1825–1908) in 1884 as organulum (plural is organula), using Latin for "small organ." He was studying protozoa and believed organelles were unique to protozoa, which lacked an internal cell structure.[1] Multicellular organisms, he believed, did not have organelles because he drew a parallel to the comparison: Organ is to multicellular organism as organelle is to single-celled organism. In 1900, Edmund Wilson and others generalized Möbius' finding to all eukaryotic cells. He referred to them as "cell organs." This left bacteria without organelles. At that time, bacteria were believed to have homogeneous protoplasm that grew in a mass and just partitioned into two by "amitotic fission." By 1920, structures like flagella, cilia, and centriole were added to the known organelles, including mitochondria, plastids, or chloroplasts (limited to plant cells), and the Golgi apparatus near the nucleus named for Camillo Golgi, who first reported it in 1892.[2]

Organelles Associated with the Optical Microscope

Richard Altmann first described mitochondria in 1889, but he used the term "bioblasts" for them.[3] Altmann also gained fame by coining the term "nucleic acid" for the nuclein that Miescher had first reported. The bioblasts were unusual to Altmann because they seemed to have their own metabolism and their own heredity. He developed a stain to bring them into more detail. These organelles were given the name mitochondria by Carl Benda (1857–1932).[4] All eukaryotic cells had mitochondria and they varied in number with the tissues. Altmann believed they were "elementary organisms" that colonized a cell and may have had a symbiotic relation to the cell. The hereditary properties of mitochondria were revived by Friedrich Meves (1868–1923), who showed that Flemming's and Benda's organelles were the same thing, mitochondria. In 1918, Paul Portier (1866–1962) saw them as having an evolutionary past, and he identified them as bacterial symbionts in cells.[5] In 1908, Carl Correns reported a case of maternal inheritance (through the ovule) in several plants.[6] These turned out to be mutations in chloroplasts, as Ruth Sager (1918–1997) discovered many years later. A similar maternal inheritance was found for mitochondria by Herschel Mitchell's (1913–2000) laboratory using the fungus *Neurospora*.[7]

In the 19th century, the cell's aster was first described and named by Walther Flemming in 1882 while he was working out mitosis. The aster is associated with the formation of the spindle in mitosis. The centrosome, within which the aster is generated, was named by Theodor Boveri in 1885. Chloroplasts in plants were identified and named by Julius von Sachs (1832–1897) in 1862.[8]

The Electron Microscope Reveals More Organelles

A major advance in cell morphology was needed to work out the structure of these organelles and to identify additional organelles in the cell. That advance was Ernst Ruska's electron microscope in 1939.[9] World War II had slowed down European contributions with that instrument, and most of the discoveries were associated with major research institutions and universities in the United States, especially Rockefeller University, which pioneered in the development of what became known as cell biology.

Albert Claude (1899–1983) was a Belgian scientist who studied at the Kaiser Wilhelm Institute in Berlin before moving in 1929 to The Rockefeller Institute for Medical Research in New York. He used an electron microscope in the late 1930s and 1940s to study cell organelles, and he also began introducing techniques in the 1930s to separate organelles from cells. He used osmotic changes to burst the cells and filtration to separate the largest fragments of cells from the soupier remnant. He placed this suspension of smaller material in centrifuges and obtained layers based on their mass. He pipetted each layer into separate tubes so he could study their chemical composition.[10] George Palade

(1912–2008) was Romanian, obtaining his M.D. in Bucharest. He heard a lecture at New York University by Claude, took an interest in electron microscopy, and moved to the Rockefeller Institute. In 1955, he described the endoplasmic reticulum and isolated ribosomes from portions of the endoplasmic reticulum. He noted the association of the Golgi apparatus with the membranes of the endoplasmic reticulum.[11]

Palade teamed up with Keith Porter (1912–1997), and they worked out the structure of the inner and outer membranes of mitochondria. Porter was Canadian, born in Nova Scotia. He worked out the 9 + 2 arrangement of filaments in flagella, cilia, and other cell filament structures, such as sperm tails.[12] The fourth major contributor to the development of cell biology was Christian de Duve (1917–2013), another Belgian, but born in England during World War I while his family lived in exile waiting for the war to end. He studied in Antwerp and Leuven when he returned and developed cell fractionation techniques to isolate additional organelles. His most noted finding was the lysosomes in 1955.[13]

The term "organelles" was still a term with conflicting definitions. Some cell biologists considered organelles to be membrane-bound units in the cell cytoplasm. This would make only the mitochondria and chloroplasts organelles. Others included any compartmentalized structure of the cell as an organelle, leaving only the diffuse "protoplasm," or cytosol, as not being an organelle. The nucleolus is not bound by a membrane. The nucleolus was observed by Schleiden and Schwann, who believed this was evidence of the free formation theory, with fluid protoplasm forming nucleoli, nucleoli forming nuclei, and nuclei forming cells. But in the 20th century, the nucleolus was noted to shrink in size during mitosis or meiosis and disappear. It reappeared almost like a Schleiden–Schwann free formation at the end of mitosis. Barbara McClintock in 1934 noted its association with a specific chromosome in maize. She called that region of the chromosome the nucleolar organizing region (NOR).[14] Later, in human cells the nucleolus was shown to be a composite of NOR regions in chromosomes 13, 14, 15, 21, and 22. Some species have multiple nucleoli, although most human cells have just one.

Cell vacuoles appear in the cell cytoplasm and often fill with water, and some discharge that water through the plasma membrane. Sometimes these vacuoles fuse with lysosomes, especially vacuoles that form around wastes, microbes, or worn-out membranous parts of other organelles. They are digested by enzymes in the lysosomes. This was largely worked out in the 1960s to 1970s. In 1966, Max Birnstiel (1933–2014) and graduate student Hugh Wallace showed nucleoli produce and store ribosomal RNA. They also isolated the first eukaryotic gene from the African clawed frog, *Xenopus laevis*.[15]

What makes the membrane-bound organelles in the cytoplasm distinctive is that both mitochondria and chloroplasts also have their own DNA. This made them candidates for an evolutionary role in the formation of eukaryotic cells from prokaryotic cells by endosymbiosis, a thesis that had early roots in the 19th century but that was put to the test by Lynn Margulis from biochemical studies and that has won over almost all cell biologists and evolutionary biologists.[16]

Cell Biology Required Tools for Studying Its Components and Functions

The techniques for studying isolated cells and components of cells vary with the instruments and procedures available at the time. In the 19th century, the major mechanical tools were agitation or use of a heated pin or a fetal hair to separate cells from one another or to obliterate a cell from a morula. Physiological techniques, including osmosis or solutions deprived of calcium ions, were used to cause cells to burst or to cause cells to lose their adhesion. In the 20th century, a variety of new approaches became possible with the use of centrifuges, chromatography, and radioactive labeling. The physical and chemical tools before the 1940s were limited to cellular, not organelle, studies. Thus in 1907, the biologist Henry Van Peters Wilson (1863–1939) used silk mesh as a means of squeezing sponges so that their isolated cells could fall into a dish. Any lumps that managed to pass through the mesh could be separated by use of pins. These disaggregated masses of cells soon swarmed to form lumps, and new sponges emerged from the lumps. When Wilson placed these sediments of cells on slides and watched them under the microscope, he saw the migration of cells and their differentiation into the specific cell types of the sponge species that he used. He showed he could also partition a sponge into two or more segments, each of which would form a full normal sponge. He also could squash two sponges together and obtain a single sponge from the commingled cells. Although this dissociation technique was effective for isolating cells, it was not effective for isolating organelles in cells to study organelle function.[17]

Wilson was typical of the new American Ph.D. that was started at Johns Hopkins University by its President, Daniel Coit Gilman (1831–1908). Students worked hard to do an original piece of research, but after leaving to join a faculty they found themselves teaching five to seven courses a year leaving little time for research. Wilson did his research mostly in the summers. Wilson got his Ph.D. with William Keith Brooks, who also had supervised the Ph.D. of T.H. Morgan and who had convinced William Bateson to switch from his interests in embryology to finding ways to shift heredity from a speculative philosophy to a science.

About a half-century later, in 1959, Slovenian Marko Zalokar (1918–2012) used different techniques to study the organelles of cells.[18] Zalokar used the fungus *Neurospora*, which produces filaments called hyphae that give the fluffy matted appearance of fungal growth on vegetation. Hyphae have a syncytial organization with no sharp membranes separating the hyphae into separate cells. Thus, they have scattered nuclei among their other cytoplasmic inclusions. He centrifuged the hyphae and found that their organelles aligned according to their masses. Fat and vacuoles were followed by a watery layer, then by nuclei followed by mitochondria, then by what were then called microsomes, and finally by a layer of starchy granular material similar to glycogen. He fed the hyphae for four minutes with a radioactive solution containing tritiated uracil. (Tritium is a radioactive form of hydrogen, and uracil is a nitrogenous base found in RNA but not in DNA.) When such labeled hyphae were centrifuged shortly after feeding and placed on sensitive photographic film, the signs of radioactivity were found in the nuclear fraction. But if the hyphae were allowed to spend an hour after radioactive feeding and then centrifuged, most of the radioactivity was found by inspection of the photographic film to be in the microsome layer. Zalokar concluded "these findings suggest that ribonucleic acid is formed in nuclei and that it migrates into the cytoplasm later." Its location in the microsomes suggested it was involved in protein synthesis, because cell biologists had previously shown microsome involvement with protein synthesis. Later, these microsomes would be analyzed in more detail, and the effective cytoplasmic component involved in protein synthesis would be called a ribosome. In 1959, protein synthesis was still in the field of cell biology. In five more years, it would be in the field of molecular biology.

Biochemistry Played a Major Role in Working Out Organelle Function

The cell organelles acquired their functions through the approaches of biochemistry, cell biology, and molecular biology. For mitochondria, the biochemistry was worked out without knowledge that this was associated with the mitochondria. In 1912, Otto Warburg (1883–1970) identified enzymatic activity in the cell associated with the uptake of oxygen.[19] This was shown in 1925 by David Keilin (1887–1963) to be associated with a series of pigmented chemicals bearing iron that he called cytochromes. Keilin described it as a chain of reactions, and the term "respiratory chain" soon became its designation.[20] He believed that electrons were transported from hydrogen and that hydrogen binding to the oxygen would lead to water formation as well as heat and other compounds formed in the cell. In 1929, Karl Lohmann identified the cytochrome respiratory chain as producing the compound ATP (adenosine triphosphate).[21] Fritz Lipmann in 1941 identified ATP as the chief energy-carrying molecule in the cell and involved with enzymes in the synthesis or degradation of virtually all other molecules involved in cell metabolism.[22] By 1950, Eugene P. Kennedy

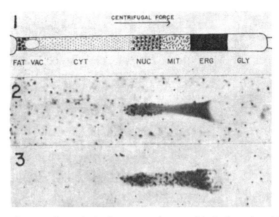

Marco Zalokar used autoradioraphy to do an experiment with the bread mold *Neurospora*. He fed the mold with radioactive uracil for one minute and then with nonradioactive uracil compounds. In the illustration on *top*, the centrifuged cell (hypha) falls into layers from lightest to heaviest: fat vacuoles, cytoplasmic fluid, nuclei, mitochondria, ergastoplasm (later called microsomes and now called ribosomes), glycogen (animal starch). After one minute, the centrifugation produces the *middle* row with the radioactivity located mostly in the nuclear faction. An hour later after being fed nonirradiated uracil, the cells show the radioactivity mostly in the ergastoplasm (what would now be the ribosomes and portions of the endoplasmic reticulum bearing the ribosomes). When reading a manuscript some 50 years after its publication, it is not uncommon to see the incremental changes that have taken place with name changes of the components and with functions added that were not known at the time. This is why the mental picture for a paradigm shift is different from the mental pictures of the evolving concept of protein synthesis as we go from Zalokar's to the present. It is not Kuhn's incommensurability of vocabulary. It is just using different, more detailed names for the same thing.

(1919–2011) and Albert Lehninger (1917–1986) identified the process of oxidative phosphorylation (the production of ATP, water, and carbon dioxide from small sugars and their breakdown products by oxygen) as the major function of mitochondria with the cytochromes serving as a component of the mitochondria. In 1967, Peter Mitchell (1920–1992) worked out the molecular events in oxidative phosphorylation and tied the chemical process to the proteins of the mitochondria with the protons (hydrogen stripped of its electron) pumped through the mitochondrial membranes.[23] Mitchell's work on the "proton pump" in mitochondrial respiration is sometimes recognized as a paradigm shift, because prior to his work biochemists recognized only electron movement among atoms as a basis for chemical reactions.

The story of oxidative phosphorylation and energy production overlapped the working out by biochemists, especially Otto Meyerhof (1884–1951) and his students, of glycolysis.[24] Oxygen is not needed in glycolysis. Glycolysis is the process, unknown to Pasteur, that digests sugars and produces fermentation. This allows for the existence of anaerobic organisms, chiefly some bacterial cells that, unlike eukaryotic cells like yeast, do not require energy from mitochondrial oxidation of sugars. Instead, the sugars are broken down by enzymes into carbon dioxide and alcohol or in some cells, acetic acid, and in other cells, lactic acid. In human tissues, muscle cells use glycolysis to produce lactic acid (eventually converted to carbon dioxide and water). Other tissues produce acetate molecules and these get recycled and eventually converted to carbon dioxide and water. Yeast cells produce alcohol as their product of glycolysis. In general, less energy and ATP are generated by breaking sugars into smaller molecules by glycolysis than by converting such sugar molecules in the process of mitochondrial oxidative phosphorylation.

The other organelles acquired their functions as biochemists and cell biologists studied them using cell fractionation, radioactive labeling, chromatography, and other tools that proliferated after World War II ended and the support of science became generous from government agencies such as the National Science Foundation and the National Institutes of Health for the life sciences. It also shifted universities to a research mode, with those investigators receiving grants teaching less. The days of a Henry Van Peters Wilson teaching seven or eight courses a year were over at research universities, allowing year-round research activities. Lysosomes were identified as having about 80 different enzymes (varying with tissue type in the organism) for digesting and recycling wastes. For humans this meant in the 1970s to 1990s the identification of numerous human genetic disorders such as Tay–Sachs syndrome (neurolipid recycling defect), Hurler syndrome (mucopolysaccharide storage in the lysosomes), or mannosidosis (carbohydrate metabolism defect). These are usually recessive disorders inherited from parents who are both carriers of the mutant alleles, and they are known as lysosomal storage disorders.

A major portion of the layers of the endoplasmic reticulum was shown to be studded with ribosomes. The ribosomes were shown to be essential for protein synthesis. The Golgi apparatus, in addition to budding off lysosomes and other vesicles, is also involved with circulating proteins throughout the cell through its connection to the endoplasmic reticulum. The plasma membrane was revealed by electron microscopy to have a double layer of lipids. The lipids are studded with proteins, many of which act as channels for the movement of molecules into or out of the cell. The lipid bilayer is the defining feature of not only the plasma membrane but also the membranes that enclose mitochondria and chloroplasts. It is also, in a doubled state like an envelope, found as the nuclear envelope. That envelope is characterized by nuclear pores through which large molecules can be passed to the cell, such as transcribed RNA from the chromosomal genes.

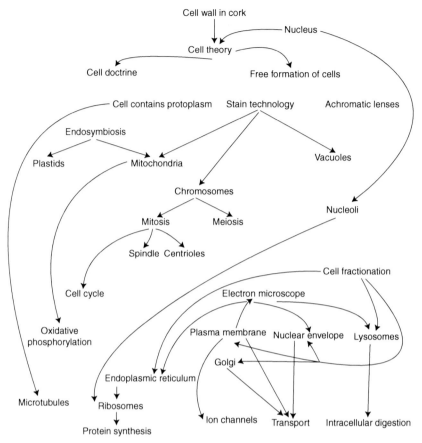

Hooke in 1665 first described cells as box-like structures in cork tissue. That protective outer layer was later (1838) identified as composed of cellulose in plants. The nucleus was described in 1835. The cell theory of 1838 filled the cell with fluid and used a free formation of cells theory as the source of new cells. By 1855, Remak and Virchow saw the cell as a living unit with all cells arising from preexisting cells. The inner fluid got the name protoplasm (displacing sarcode). Stain technology and achromatic lenses in the late 1850s allowed greater detail to be studied and named. Vacuoles, Golgi apparatus, and mitochondria were named. In the 1870s, the shift from histology to cytology occurred as interest focused on the nucleus and its role in cell division, called mitosis in body (somatic) cells and meiosis in germinal tissue. The components of the spindle apparatus, centrosome, and centrioles were worked out in the 1880s. The cell organelles became more detailed and new ones were added after the introduction of the electron microscope in the late 1930s and the introduction of cell fractionation and chromatography in the 1950s. Mitochondria were associated with oxidative phosphorylation. The cell membrane had numerous ion channels that allowed molecules and ions to pass into or out of the cell. The endoplasmic reticulum was a spongy network studded with ribosomes, which were identified as playing a major role in protein synthesis in the 1970s. Lysosomes were shown to be organelles for digestion of wastes and damaged organelle components. The stages of mitosis were functionally associated with the cell cycle, whose physiology and biochemistry were worked out.

Cell Biology Is Largely Based on Incremental Events

The study of organelles was largely a 20th century activity. It begins with 19th century morphology and quickly shifts to biochemistry and electron microcopy in the mid-20th century. It shifted to cell biology with its fractionation techniques, and shifted again to a more detailed biochemistry and molecular biology in the last third of the 20th century. As in other branches of biology, the study of organelles is characterized by a growing reductionist view of the living cell. Its components can be isolated and their functions worked out. There is no Kuhnian paradigm shift of a new major theoretical way of looking at cell components. The shifts are mostly produced by new tools and not new theories, with relatively few exceptions like Mitchell's proton pump (chemiosmotic) model of oxidative phosphorylation, which some cell biologists regard as a paradigm shift. The new theories that do occur emerge from the new data and not a shuffling and renaming of the components of the past. The image in the 19th century of the cell as the basic unit of life is still present, but the contents have been identified through new technology and their functions worked out by fusion of descriptive biology with new emerging fields of biochemistry, cell biology, and molecular biology.

References and Notes

1. Möbius KA. 1884. Das Sterben der einzelligen und der vielzelligen Tiere. Vergleichend betrachtet. *Biol Zentbl* **4**: 389–392, 448.

2. Möbius, op cit.

3. Altmann R. 1889. *Die Elementaorganismen und ihre Beziehungen zu den Zellen*, p. 145. Veit and Co., Leipzig, Germany.

4. Benda C. 1898. Weitere Mitteilungen über die Mitochondria. *Verh Physiol Gest Berlin Jahrb 1898/1899* **4–7**: 376–383.

5. Portier P. 1918. *Les Symbiotes*. Masson, Paris.

6. Correns C. 1908. Vererbungs versuche mit blass (gelb) und bündtblättrigen Sippen bei *Mirabilis, Urtica*, und *Lunaria*. *Z Indukt Abstammungs Vererbungsl* **1**: 291–329.

7. Mitchell MK, Mitchell HK. 1952. A case of maternal inheritance in *Neurospora crassa*. *Proc Natl Acad Sci* **38**: 442–449.

8. Wilson EB. 1896. *The cell in inheritance and development*. Macmillan, New York.

9. Ruska E. 1986. The development of the electron microscope and of electron microscopy. In *Nobel lectures in physics 1981–1990* (ed. Ekspång G). 1993. World Scientific Publishers, Singapore.

10. Claude A. 1943. The constitution of the cytoplasm. *Science* **97**: 451–456.

11. Sabatini PD. 1999. George E. Palade: Charting the secretory pathway. *Trends Cell Biol* **9**: 43–47.

12. Porter KR, Claude A, Fullam EF. 1945. A study of tissue culture cells by electron microscopy. *J Exp Med* **81**: 233–246.

13. De Duve C. 1984. *A guided tour of the living cell*. Scientific American Books, W.H. Freeman, New York.

14. McClintock B. 1938. The fusion of broken ends of sister half chromatids following chromatid breakage at meiotic anaphase. *Miss U Agr Exp St Bull* **290**: 1–48.

15. Birnsteil ML, Wallace H, Sirkin JL, Fischberg M. 1966. Localization of the ribosomal DNA components in the nucleolar organizing region of *Xenopus laevis*. *Natl Cancer Inst Mono* **23**: 431–449.

16. Margulis L. 1970. *On the origin of eukaryotic cells*. Yale University Press, New Haven, CT.

17. Wilson HV. 1907. On some phenomena of coalescence and regeneration in sponges. *J Exp Zool* **5**: 245–258.

18. Zalokar M. 1957. Nuclear origin of ribonucleic acid. *Nature* **183**: 1330.

19. Warburg O. 1911. Über die Rolle des Eisens in der Atmung des Seeigeleis nebst Bemerkungen über einige durch Eisen beschleunigte Oxydationen *m. Abb. Sitzungsber Heidelberger Akad Wiss Math-Nat Kl B Heidelberg, Germany*. Academy of Sciences, Heidelberg.

20. Keilin D. 1925. On cytochrome, a respiratory pigment, common to animals, yeast, and higher plants. *Proc Royal Soc Biol Sci* **98**: 312–339.

21. Lohmann K. 1929. Über die Phosphorophatfraktion im Muskel. *Naturwissernschaften* **17**: 624–625.

22. Lipmann F. 1941. Metabolic generation and utilization of phosphate bond energy. *Adv Enzym* **1**: 99–162.

23. Mitchell P. 1966. Chemiosmotic coupling in oxidative and photosynthetic phosphorylation. *Biol Rev Camb Philos Soc* **41**: 445–502.

24. Meyerhoff O, Oesper P. 1947. The mechanisms of the oxidative reactions in fermentation. *J Biol Chem* **170**: 1–22.

CHAPTER 12

Evolution: From Guesswork to Natural Selection to Molecular Phylogeny

Fixity of species versus varietal changes, sterility of interspecies hybrids, domestic selection and new breeds, evolution as environmentally induced (Lamarck). Darwinian natural selection, evolution by polyploidy, evolution by aneuploidy, chromosome rearrangements as isolating factors, microevolution versus macroevolution, evolution and ecology, Darwinian gradualism versus punctuated equilibrium. Attempts to show evolution by experiments, evolution and phylogenic relations from the fossil record, phylogenetic relations by comparative embryology, evolution and phylogeny by protein sequences, evolution and phylogeny by comparative genomics.

The idea of an evolution of life was present in the thinking of Greek philosophers more than 2000 years ago. This is not a surprise, because all humans must have been curious about how they and other life came to exist on earth. There are two major models to consider. One is an act of creation by a Creator, which is supernatural and usually associated with a religious tradition. That was not a problem for the Greeks, because even their polytheistic deities were created by preexisting gods for several generations of gods. It is just a mental way to shunt a difficult problem to a more remote stage (often eternal) and not think about it much. The alternative to this is an evolution from nonlife. Here, too, there may be different ideas on how this occurs. Some Greek philosophers speculated that there was a chaos of parts trying to fit together. Most of these adhesions were not functional, and only the ones that put the pieces in a proper working arrangement survived and proliferated.[1] It is a kind of natural selection by default. An ideal working organism needs all its components in their appropriate places. Other Greek philosophers believed in a spontaneous generation of life, at least for most forms below the vertebrates and including some mammals like mice that arose out of trash stored for long periods of time or other life from rotted meat that generated maggots.

Modern experimental science, already noted, arose in the Renaissance with Galileo pioneering the physical sciences and Harvey pioneering the life sciences. Although Harvey showed the circulation of the blood by experimental means, his work offered no challenge to religious orthodoxy. Nor did Vesalius's detailed anatomical drawings of the body's bones, muscles, and nerves prove disturbing to religious authorities (other than the possible moral aspects of dissecting human cadavers). Galileo, in attempting to show Copernicus' solar system model of the earth in relation to the planets, sun, and stars, was more dangerous to the religious traditions of his day because there were both scriptural and theological reasons why that model was a threat to religion. It shifted the motions of heavenly bodies to science and mathematics. It made material a moon with mountains and craters. It made material a planet, Venus, that had phases like the moon. It showed imperfections in the Sun, especially sunspots that could be used to show the Sun rotated. It revealed the presence of four moons around Jupiter.

Biblical accounts of creation in Genesis put the Earth's origin on the third day, after light and dark (the first day), and waters above and below a firmament (the second day). On the third day, plant life, including grasses and fruit trees, also appears on the first appearance of land. On the fourth day, the Sun, moon, planets, and stars are created. Animals of the sea and animals in the air are the activities of creation for the fifth day. Land animals and humans are the product of the sixth day. The seventh day was rest. Adam and Eve are described in the second account of creation.[2] For biologists, the creation accounts suggested a sexual dualism of male and female for all living things (reinforced by the story of Noah's ark and the two-by-two entry on the ark's gangplank). It also suggested fixity of type (the term species was unknown in biblical times, with the phrase "each after its own kind" as the suggestion of fixity). It also suggested a capacity for reproduction to fill the niches of the Earth. Although there was not much biology to serve as a text in Genesis, there were interpretations that served as impediments to a scientific study of life. Those clashes began when the expanding colonization of the world brought back to Europe massive numbers of new plants and animals from Africa, Asia, and the New World.

Classification before Linnaeus

The arrival of new plants and animals in Europe led to methodical ways of classification. Before the 16th century, botany was virtually identical to herbal medicine and was a branch of medicine. The first attempt to describe plants without reference to their medicinal values was by Otto Brunfels (1488/1489–1534), who shifted from being a Catholic monk to a Protestant during the Reformation and took up botany as his concentration of interest. He described and gave the German names of the plants found in Germany with careful illustrations he prepared for his Latin text of 1530.[3] A more thorough classification of plants appeared in 1640 with the publication of Caspar Bauhin's (1560–1624) study of the 5640 known plants at that time. Bauhin named and introduced the idea of species and of a collection of species or a genus. He also used a genus and species two-name designation, but his method of classifying plants was not consistent for the varied criteria he used.[4]

The most effective work in classification, before Linnaeus, was published by Joseph Pitton de Tournefort (1656–1708). He had traveled to the Far East and collected 1356 new species. He was the first to conceive a category that related several species. He called this a genus, using Bauhin's terminology. His description of each species was laborious with seven to 12 Latin descriptors making it too cumbersome to memorize.[5] The difficulty with all of Linnaeus' predecessors was they lacked a consistent way to compare plants or animals. Describing thousands of different species without a common reference might fill a volume with data but it did not organize plants or animals in any systematic way.

Linnaeus Provides a Consistent Classification

Carl Linnaeus introduced an effective means of classification by using plant or animal anatomy as his principal method. For plants, it was sexuality or, more precisely, the floral structures, including the number and location of stamens and pistils and inflorescent organization.[6] This gave Linnaeus a means of generating relatedness beyond the genus level. For animals, he chose anatomy, including radial versus bilateral symmetry, vertebrate versus nonvertebrate dorsal structure, and external versus internal skeleton. This systematic method of classification grouped animals into rational clusters. Linnaeus also included humans among the primates. He used the binomial method of Tournefort modified from Bauhin and added families, orders, and classes. This was open-ended and allowed for additional levels of organization. Linnaeus published his first edition of *Systema Naturae* in 1735.[7] It went through 12 editions under Linnaeus; the 13th edition was published in 1770 and totaled 3000 pages. Linnaeus made use of his own expeditions and those of his students who traveled with explorers to bring back new plants and animals.

Linnaeus created several problems for his contemporaries. For the pious, there was the discomfort of lumping humans as *Homo sapiens* with apes and monkeys. Linnaeus was neither a deist nor an atheist. He abided by his Lutheran faith, but he insisted that humans had to apply the same criteria of classification to themselves as they did to other animals, and anatomy was essential for that classification. More important for scientists in the life sciences, Linnaeus forced them to explain why these groups existed. For many, this was another way to read the mind of God, whereas for others it was a way to seek relationships in what would soon become comparative anatomy. The organized collection of anatomical features also had functional implications about the types of lives they lived,

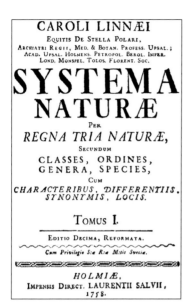

CAROLI LINNÆI
Equitis De Stella Polari,
Archiatri Regii, Med. & Botan. Profess. Upsal.;
Acad. Upsal. Holmens. Petropol. Berol. Imper.
Lond. Monspel. Tolos. Florent. Soc.

**SYSTEMA
NATURÆ**
Per
REGNA TRIA NATURÆ,
Secundum
CLASSES, ORDINES,
GENERA, SPECIES,
Cum
*CHARACTERIBUS, DIFFERENTIIS,
SYNONYMIS, LOCIS.*

Tomus I.

Editio Decima, Reformata.

Cum Privilegio Sia Ria Mitis Svecia.

HOLMIÆ,
Impensis Direct. LAURENTII SALVII,
1758.

Carl Linnaeus was a Swedish botanist, who taught at Uppsala University. In his early career, he explored the Lapland area of northern Sweden and did an inventory of the animal and plant life he found. He made four such expeditions and published his work, attracting several students to do expeditions on voyages to Asia and the New World. As his fame increased, so did donations of specimens, drawings, and descriptions. He amassed this information and used a binomial nomenclature as his method of classification—a genus name followed by a species name, both in Latin. Thus, you and I are of the species *Homo sapiens*. It was an improvement over all prior attempts at classification, being both more inclusive and more scholarly. His first efforts were in botany. Anatomy and body plan were the most important features of his classification. Linnaeus was not rigid on the fixity of species and believed hybrids could lead to new species.

food they ate, and conditions in which they could find habitats. Finally, Linnaeus' method of taxonomy created the issue of stability. Why did no two people look exactly alike? Why was this true of almost all individuals of a species? He still remembered his shock among the Lapps near the Arctic Circle when they could identify and name each member of their herds.[8] What was the cause of that variation? Why did some species of plants that he mated form hybrids with all its members sterile, whereas other crosses yielded hybrids with partial or even full fertility? How fixed were species if they were collected from different continents?

Goethe Introduces a Holistic "Nature Philosophy"

The first application of Linnaean taxonomy was not to evolution but to a theory of archetypes suggested by Johann Wolfgang von Goethe (1749–1832). Goethe was a multitalented person with an education in mining engineering and a gift for the humanities as Germany's greatest writer, poet, novelist, and playwright. His greatest creation, *Faust*, reflected his personality. His curiosity was robust, and he made contributions to botany, zoology, anatomy, and color theory. Goethe was also a founder of the Romantic Movement, which tended to embrace vitalism and holism and to reject reductionism as inadequate. He believed he could make sense of Linnaeus' methods of classification by a Platonic approach. Plato saw the ideal as "real" and the specific forms of that ideal as ephemeral. To Goethe this meant there had to be an ideal plant, an original archetypal plant (perhaps in the mind of God) from which all other plants arose.[9] This "*Ur*" (the German word for original) plant, or *Urpflanze*, consisted of a single root, stem, and single leaf. All other plants would be variations of the *Ur* plant. This allowed God to create the *Ur* plants and *Ur* animals, and the variations since then would be a consequence of their responses to their environments and the demands put on them by their habitats as they moved away from their site of creation. One implication of Goethe's *Ur* life was the unlikely idea of a fixity of species. In *Geschichte meines botanischen Studiums* (*The Story of My Botanical Studies*, 1830 or 1831), he wrote "they [plant forms] have been given a felicitous mobility and plasticity that allows them to grow and adjust themselves to many different conditions in many different places."

Scientists throughout the last half of the 18th century and the first half of the 19th century used Linnaeus to bolster their views of the fixity of species or to deny such fixity. Linnaeus was also ambivalent in his lifetime and believed some hybrids could lead to new species formation. Compounding the diversity of beliefs about species were the numerous breeders creating startlingly different sizes, shapes, colors, and distortions of the features of the dogs, cats, song sparrows, finches, budgerigars, and livestock they enjoyed raising for profit or hobby.

Darwin's Voyage around the World Shaped His Thinking on the Species Problem

It was this background that shaped the young Charles Darwin before he set sail on his voyage around the world as a ship's naturalist.[10] He knew that Linnaeus, despite his occasional doubts, favored fixed species, as did virtually all of Darwin's professors at Cambridge. He knew from his grandfather's work and Jean-Baptiste Lamarck's writings at the start of the century that a theory proposed by Lamarck of the inheritance of acquired characteristics rejected the fixity of species. To Darwin, Lamarck and those who cited Linnaeus were both basing their views not on evidence but on theory. Darwin made no commitment going forth on his voyage in 1831, but by the time he returned five years later, his view of the fixity of species was shattered. Six years after his return, in 1842, he felt confident enough to write his first sketch for a theory of natural selection leading to the formation of new species.

Darwin's transformation came from his observations that animals on islands off Africa's west coast resembled animals that were present in West African countries of similar latitude. This was also true for animals he observed in the Galápagos Islands off the west coast of South America. Yet the islands in both oceans were volcanic and had similar climates. Something other than a simple common environment was causing variations in species to shift to new species or even genera on these islands. A second line of evidence struck Darwin when he looked at the animals that were present on the north shore and the south shore of the Amazon River. They differed. But the birds that flew no more than a mile to get across the river were identical. The river served as a barrier to the land animals but not to the birds. A third line of evidence came when he was in Patagonia and studied some sloths. He found that in the soil were fossilized giant-sized sloths that were now extinct. This suggested that sloths had changed over long periods of time residing in the same land. He was puzzled by the abundance of marsupial forms in Australia and their rarity in other continents.[11] What was the cause of the biodiversity in relation to geography? Why were some Linnaean taxa rare on islands, like amphibians, whereas lizards and snakes were not?

When Darwin returned, he first wrote up his narrative of the *Beagle* voyage. This was a best seller as both a travel book and a detailed account of a naturalist with an eye for detail and a capacity to raise questions about the significance of his observations. He did not propose a theory of evolution. He documented the evidence that could be used for such a theory that had been amassing in his five years of explorations. It is hard not to read the *Voyage of the Beagle* today without

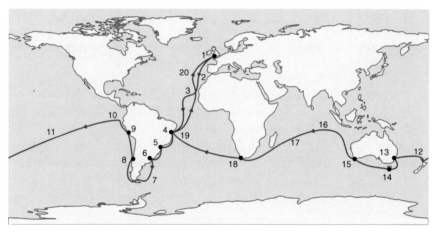

Darwin's round-the-world trip took place on the *HMS Beagle* under Captain Robert FitzRoy. The ship left England in 1831 and did not return until 1836. Most of Darwin's studies of plant and animal life were gleaned from his visits to coastal South America and the islands near Africa (*3*, Cape Verde) and the eastern coast of South America (*7*, Falkland Islands). But the most informative island life was in the Galápagos Islands (*10*). Quite distinct were the marsupial animals he saw in New Zealand (*12*) and Australia (*13, 14, 15*). Equally revealing were his studies of Brazilian and Argentinean present-day and fossil life (*4, 5, 6*). Darwin took voluminous notes and wrote letters while at sea that he sent to his mentor John Henslow, who had recommended him for the voyage. In 1839, Darwin published *The Voyage of the Beagle*, which became an international bestseller. His theory of evolution would wait another 20 years when published as his book *On the Origin of Species*.

seeing the evolutionary implications jumping out of the pages. According to Darwin, the idea came to him in 1838 and he realized it would require enormous documentation. The idea was simple. Just as breeders select for a type and "fix" a breed of dog or roses, so too does nature fix a species by selection. The breeder uses conscious selection with a purpose of selling a new breed or selling a popular breed that has to be carefully kept free of "rogue" variations. Nature uses an unconscious selection for all living things that sifts variations most adapted to their environments and habitats. He called that unconscious selection "natural selection." This meant he had to do a detailed study of breeders of plants and animals and establish how they did this and how long it took to fix a breed. He also wanted to know the source of the variations. Did this get fixed quickly with new variations showing up sporadically as "sports" or did they arise as subtle departures that over many generations accumulated in the breed?

Darwin also wanted to study the details of the species of a single group of animals, and he chose the barnacles to do this. He became so adept at identifying them that he could tell within about 50 miles the location of any barnacle given to him. He also embarked on experiments. Could all of the life of the Galápagos have come from the shores of western South America? He soaked branches with seeds in salt water for the amount of time ship captains told him it would take for a storm to send them drifting toward the Galápagos from Ecuador or Columbia. He showed some seeds could survive such a saline drenching.

Darwin attributed to Thomas Malthus (1766–1834) the impetus for his theory of natural selection. Darwin read Malthus' book on population and applied its findings to nature. Organisms are kept in a balance between their exponential growth rate and their resources. If they exceed that balance, animals and plants, just like people, do not survive.[12]

Darwin's Care Not to Rush into Print with His Theory of Evolution

Chapter by chapter, Darwin folded in the findings from his readings, his correspondence with other naturalists, and his own data. He knew that his theory would lead to controversy (he said it made him feel like a murderer). If it were just a scientific controversy, that was tolerable. But it would also offend many pious people whether they were scientists or not. There was nothing Darwin observed, studied, or eventually wrote down in his notebooks or his published writing that suggested life was created by God or that God was directing the path of speciation toward some ultimate good, or that the process of evolution in some way could reflect the benevolence of God. He had not read the mind of God. He had not read the Bible of nature. He had worked out a scientific theory of natural selection that brought about an evolution of life, an actual account of the "origin of species."

In 1858, Darwin was forced to publish his work in progress when he received a copy of Alfred Russel Wallace's (1823–1913) manuscript that contained an identical account of evolution by natural selection.[13] Darwin's *On the Origin of Species* appeared in 1859 and the debates were numerous as Darwin anticipated.[14] He kept out of them and allowed his friends, Joseph Dalton Hooker (1817–1911) and Thomas Henry Huxley, "Darwin's bulldog," in England and Harvard botanist Asa Gray (1810–1888) in the United States, to respond to the criticisms. Year by year, scientists joined the Darwinian view and they added new evidence and expanded Darwin's findings and theory. The fossil record was meager when Darwin published in 1859, but soon expeditions were digging up fossil mammals and reptiles to reveal a world of extinct animals that could not be attributed to a Noachian flood. Soon those fossils would include human skeletal remnants that could be distinguished as a different species of *Homo*. The Neanderthals were the first to be unearthed. A skull of *Pithecanthropus erectus* (later *Homo erectus*) came later in the 19th century. Many of the fossils revealed the stages of transition among the taxa of cognate families, orders, classes, or even phyla. The geological record was also expanding the type of events characteristic of the rocks excavated and the fossils embedded in them. A rough dating began to appear, greatly augmented in the mid-20th century with developments in physics identifying radioactive isotopes that could be used to date the rocks with precision.

Hugo de Vries Challenges Darwinian Selection by Gradual Change

At the biological level, Darwinism encountered rival theories of evolution and experimental findings of evolution in action. Hugo de Vries claimed that evening primroses (*Oenothera lamarckiana*), being ancestral, gave rise to new species of *Oenothera* in sudden jumps (saltations) similar to the "sports" Darwin discussed with breeders in the 1840s. The "mutation theory" of evolution that de Vries proposed turned out to be false.[15] His new species were polyploids, complex aneuploids, or clusters of gene expressions created by recombination from sets of translocated chromosomes in the original *O. lamarckiana*. Those mechanisms were a rarity in both the plant and animal world. de Vries believed species and varieties were discontinuous in their origin. Opposed to his saltationist interpretation were Morgan and his students using the chromosome theory of heredity and the theory of the gene to advocate compatibility between Darwinian gradualism and the assorted chief genes and modifiers that made subtle differences in wing shape possible for fruit flies. The fly lab model of classical genetics was soon wedded to Darwinian evolution along with cytogenetic findings in fruit fly species that showed genetic isolating mechanisms leading to hybrid sterility

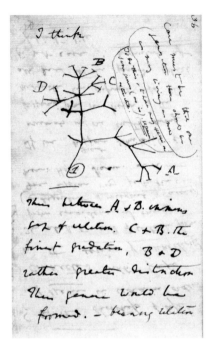

Darwin was the first to sketch a phylogenetic tree as a suggestion of showing relatedness (*left* in this 1837 sketch from his "First Notebook on Transmutation of Species"). The handwritten text is: "I think case must be that one generation should have as many living as now. To do this and to have as many species in same genus (as is) requires extinction. Thus, between A+B the immense gap of relation. C+B the finest gradation. B+D rather greater distinction. Thus, genera would be formed. Bearing relation" (next page begins) "to ancient types with several extinct forms." The contemporary tree (*below*) shows three domains of life. On the *left* are bacteria, in the *middle* are Archaea. They are both single-celled prokaryotes. On the right are the eukaryotes ending with three twigs—animals, fungi, and plants as the most recent forms of eukaryotic life.

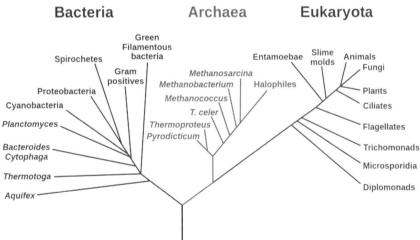

between species. These isolating mechanisms included translocations and inversions studied by many biologists using cytogenetics and collections obtained in extensive field trips,[16] including John T. Patterson (1878–1960) and Wilson S. Stone (1907–1968) at the University of Texas and by Alfred Sturtevant and Theodosius Dobzhansky (1900–1975) at Caltech. Dobzhansky later moved to

Columbia University and published a powerful case for such cytogenetic isolating mechanisms in his 1937 *Genetics and the Origin of Species.*[17]

Ecology Emerges with Unresolved Conflicts of Holism and Reductionism among Scientists and a Conflict between Scientific and Political Priorities

The term "ecology" was first coined by Ernst Haeckel in 1869 to represent the growing recognition that life on Earth was connected through evolution, geology, and geographic distribution. The roots of ecology come from studies of voyages taken by Carl Linnaeus in Sweden to Lapland, by naturalist and explorer Alexander von Humboldt (1769–1859) in his trips in the Andes where he followed plant and animal distribution by elevation and its effect on climate, and by the theory of evolution by natural selection developed by Charles Darwin and independently by Alfred Russel Wallace. Geologists, especially Charles Lyell (1797–1875), provided abundant evidence that the history of Earth's crust has changed over time. The major geological eras were worked out in the mid-19th century.

Ecology also introduced the idea of a biosphere and the conditions that permit life to exist. By 1900, Henry Chandler Cowles (1869–1939) described

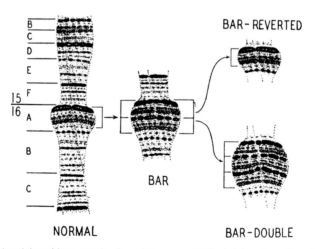

NORMAL   BAR-DOUBLE

In 1936, both Calvin Bridges at Caltech and Hermann Muller's laboratory, then in the USSR, published articles that showed the bar-eyed mutation in fruit flies was a physical duplication demonstrable by cytological study of the giant salivary chromosome that T.S. Painter had described two years earlier. Bridges discussed his finding in terms of position effect in relation to dose and juxtaposition. Muller described his finding as a demonstration that all genes come from preexisting genes or what can be called a "gene doctrine." Muller and Bridges also both argued that tandem duplications were of evolutionary importance for the gradual change in function of the two genes.

ecological succession when disasters such as fires, droughts, and floods led to massive loss of animal and plant life. Ecologists have struggled to understand the food chains and webs of relatedness among living things in an ecosystem. Most believe the complexity can be resolved through sufficient research. Others believe that the complexity defies reductionism and ecological systems are holistic. Among the more extreme holistic views are those of James Lovelock (b. 1919) who claimed the Earth was a single complex organism called Gaia. Most evolutionary biologists and ecologists reject this as an aesthetic response to complexity verging on religious feeling. Independent of this philosophic

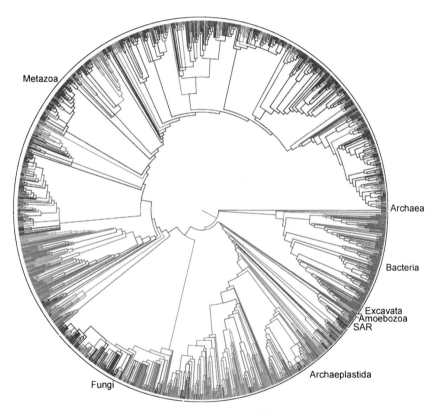

Darwin's phylogenetic tree was a small branching set of lines. At the start of the 20th century, phylogenetic trees resembled branching trees. The combination of taxonomy by studying DNA genomes and vastly increased efforts to collect and identify new species by naturalists in the last half of the 20th century have resulted in circular clock-faced phylogenetic trees. We read this one from 3 o'clock going clockwise from Archaea (which are prokaryotes) to bacteria (also prokaryotes) at about 4 o'clock, then to protozoa and single-celled eukaryotic algae between 5 and 6 o'clock. Fungi arise at 7 o'clock, and higher plants (a branch of the Archaeplastida) appear at about 6 o'clock. Multicellular animals (metazoans) appear about 10 o'clock.

conflict are the many conflicts between ecologists who raise concerns about wildlife management, ecological destruction of environments, and, in these early decades of the 21st century, the credibility of the science involved in global warming and other aspects of severe climate change around the world. It is not a holism versus reductionism debate. It is a debate over science and politics.

The New Synthesis Fuses Genetics with Evolution by Natural Selection

In the 1940s, the fusion of classical genetics and evolution by natural selection was called "the new synthesis" or "the modern synthesis." It became the prevailing model for the rest of the 20th century and is still the dominant model of Darwinian evolution in the early 21st century. It included contributions by

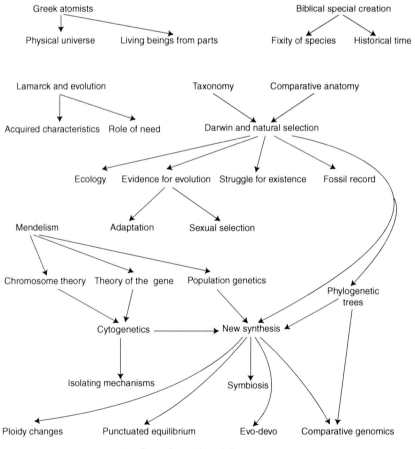

See figure legend on following page.

Theodosius Dobzhansky in 1937, Julian Huxley in 1942, George Gaylord Simpson in 1944, and Ernst Mayr in 1959.[18] In 1975, E.O. Wilson added sociobiology. Added to the new synthesis is the role of endosymbiosis leading to the evolution of eukaryotic cells from prokaryotic cells, especially the work of Lynn Margulis. Also added as the 20th century drew to its close was the work initiated by Edward Lewis in 1978 using homeotic mutations in the bithorax series of fruit flies to establish a relation between the pseudoallelic sequence of those components of the bithorax region to the corresponding segments of the arthropod body plan during its metamorphoses into an adult thorax and abdomen with wings, legs, and halteres. Lewis' model led to the discovery of homeobox regions, usually in duplicated sets of four, as widespread throughout the eukaryotic animals in bringing about anterior to posterior segment differentiation. The new field of embryology fused with the new synthesis became known as "evo-devo."

One departure from Darwinian gradualism was the proposal by Stephen Jay Gould (1941–2002) and Niles Eldredge (b. 1943) that challenged the tempo of evolution.[19] Based largely on the rapid speciation of cichlid fish in African lakes, Gould and Eldredge argued that speciation is sudden and occurs in spurts. In this sense it resembles de Vries' mutation model, which also argued for "mutating

---

Greek philosophers like Democritus proposed an atomic theory in which the observable universe was composed of atoms. Lucretius accepted this view and tried to explain the origin of life as coming from a chaotic mixture of body parts that randomly came together, mostly ineffective, but once an organism had all its parts it could reproduce and establish a growing presence on Earth. When the Bible was written, Hebrew scholars argued that the universe was created in six days and that this occurred in historical time (about 6000 years before the Bible was written). Evolution was introduced by Lamarck, who was struck by the fossil specimens in Parisian collections and by the diversity of life he observed as an officer in the French army. He believed acquired characteristics were inherited and that an organism's need (French *"besoin"*) led to speciation (such as giraffe necks and a food source). As the fields of taxonomy and comparative anatomy developed, they influenced Darwin to shift evolution not to need but to natural selection. He proposed overpopulation, limited resources, a struggle for existence, and the survival of the fittest as hallmarks of his theory. He introduced sexual selection as a lesser force. Mendelism arose independently of evolutionary studies. But hybrids were a puzzle in the mid-19th century. They were often sterile and distinctions between species and varieties that were hard to demarcate. Mendelism led to population genetics and cytogenetics, all incorporated into classical genetics. Ecology also arose from the fusion of geology, geography, and evolution. These merged with the evidences of evolution of Darwinism and became the "new synthesis" in the 1940s. Punctuated equilibrium challenged the relatively minor changes each generation that Darwinists favored. Lynn Margulis' theory of symbiotic association for organelle origins extended to the origins of higher categories of taxonomic body plans. The new synthesis has now been supplemented with evo-devo and comparative genomics at the molecular level.

periods" when a sudden rush of new species emerged. There is no cytological counterpart, however, in the cichlid evolution story, and geneticists are still divided on how many generations are involved in fixing new species in these lakes over the last few hundred thousand years. Gould and Eldredge called their model of evolution "punctuated equilibrium."

The most recent fusion of evolution has been its union with molecular biology, especially the sequencing of macromolecules and the application of large-scale sequencing to whole-genome sequencing. This made possible a field of molecular phylogeny.[20] The sequences of nucleotides and number of genes and their functions can now be used to construct elaborate ancestries of the species to other cognate taxa and chase these down through the full 92 phyla of animals and plants and even down to prokaryotic levels. The evidence so far is consistent with a new synthesis model in which the speciation process shows the roles of individual gene mutations (many of them referred to as SNPs, or single-nucleotide polymorphisms), gene duplications, small inversions, larger inversions and translocations, deletions, and various transposable and repeating small sequences of DNA as the major mechanisms of speciation. Polyploidy and aneuploidy play a more significant role in the formation of cognate taxa. Homeotic mutants also play a major role in the changes in body plans associated with different taxa. There also appears to be more horizontal gene transfer from remarkably different species. This occurs in rotifers and tardigrades in which sexual reproduction is rare or absent.[20]

If this presentation of the fusion of fields since 1859 is correct, Darwinian evolution has progressed through changes consistent with incrementalism and not with the sudden paradigm shifts in theory that have replaced it. It is not just in name only that evolution is still thought to be Darwinism. The model of natural selection has survived more than a century and a half of challenges.

References and Notes

1. Osborn HF. 1894. *From the Greeks to Darwin: An outline of the development of the history of the evolution idea.* Macmillan, New York.

2. Genesis 1.

3. Brunfels O. 1530–1536. *Herbarum Vivae Eicones.* Illus Weiditz H. Apud Joannem Schottem, Argentorati.

4. Bauhin C. 1671. *Pinax Theatri Botanici.* Impensis Joanna Regis, Basel, Switzerland.

5. De Tournefort J. 1700. *Eléments de Botaniques, ou Méthode pour reconnaître les Plantes.* Typographia regia, Paris.

6. Linnaeus C. 1737. *Genera Plantarum.* Wishoff et Wishoff, Leyden, Netherlands.

7. Linnaeus C. 1735. *Systema Naturae.* Haak, Leyden, Netherlands.

8. Blunt W. 2004. *Linnaeus: The compleat naturalist*. Lincoln, London.

9. Goethe W. 1952. *Goethe's botanical writings* (trans. Mueller B). Oxbow Press, Woodbridge, CT. Also see Goethe's *Story of my botanical studies* (1831); quote attributed by Teichmann F (tran. McAlice J). 2005. The emergence of the idea of evolution in the time of Goethe. *Res Bull* **11**: 1–9.

10. Browne J. 1995. *Darwin voyaging*. Cape, London.

11. Darwin C. 1838. *Voyage of the Beagle*. Murray, London.

12. Malthus T. 1798. *An essay on population*. Johnson, London.

13. Wallace AR. 1858. Darwin papers to the Linnaean Society. London.

14. Darwin C. 1859. *On the origin of species by means of natural selection, or the preservation of favoured races in the struggle for life*. Murray, London.

15. Carlson E. 1966. *The gene: A critical history*. Saunders, Philadelphia.

16. Patterson JT, Stone WS. 1952. *Evolution in the genus* Drosophila. Macmillan, New York; http://wwwgeneticsorg/content/157/1/1/. Wagner RP, Crow JF. 2001. The other fly room: J.T. Patterson and Texas genetics. *Genetics* **157**: 1–5.

17. Dobzhansky T. 1937. *Genetics and the origin of species*. Columbia University Press, New York.

18. Huxley J. 1943. *Evolution: The modern synthesis*. Harper, New York.

19. Gould SJ, Eldredge N. 1972. Punctuated equilibrium: An alternative to phyletic gradualism. In *Models in paleontology* (ed Schopf TJ), pp. 82–115. Freeman Cooper, San Francisco.

20. Yoshida Y, Koutsovoulos G, Laetsch DR, Stevens L, Kumar S, Horikawa DD, Ishino K, Komine S, Kunieda T, Tomita M et al. 2017. Comparative genomes of *Hypsibius dujardani* and *Ramazziottius varieornatus*. *PLoS Biol* **15**: 2002266. Welch D, Meselson M 2000. Evidence for the evolution of Bdelloid rotifers without sexual reproduction or genetic exchange. *Science* **288**: 1211–1215.

How Does Science Usually Work?

The rarity of sudden paradigm shifts in the life sciences. The overwhelming evidence that the basic concepts of biology have evolved incrementally, how incrementalism and pruning work, the necessity of new technologies for major changes in concepts. The limits of theory. The profound role of experimentation to supplement descriptive biology.

There are, I believe, five major concepts or broad theories for the life sciences.

1. The *cell theory*, which proposes that the smallest independent living unit is the cell and organisms are composed of single cells or communities of cells.

2. The *theory of the gene*, which proposes that all life emanates from the capacity of genes to make cell components and to evolve through mutations in those genes.

3. The *theory of the life cycle*, which claims that all organisms have a developmental cycle through a series of stages from fertilization, birth, or emergence to death (or for single-celled organisms through the stages of their cell cycle).

4. The *theory of evolution by natural selection*, which claims that all present life on earth had its origin from ancestors that were modified going back to the first cellular life or its origin from organic chemicals as the first replicating gene-like system.

5. The *theory of life as a molecular system*, in which living matter can be analyzed by physical and chemical means and its components isolated and reconstituted to reveal how cells and their components are organized and function without recourse to supernatural explanations.

Incremental Change Is the Basis of Life Science Change

It is my belief that all five of these broad interpretations of life are consistent with a reductionist outlook of science. They presently reject vitalism, holism, or supernaturally directed or maintained influence for their existence or functioning. Each theory has a history of incremental change as evidence for the theory

169

accumulated and as it was modified through new technologies and by experimentation. Each theory had its debates over how it worked and whether or not some nonmaterialistic explanation was possible. Not only did the science evolve, so did the cultural values. A medieval view would be highly holistic, vitalistic, and supernatural in almost all its sciences and certainly in the life sciences discussed here. Astrology, alchemy, numerology, original sin, other worldliness, and purpose or goal direction were pervasive in interpreting life and its significance. In the Renaissance, those medieval values began to yield to the new descriptive sciences and the introduction of experimental science and new tools for amplifying the human senses. But at the same time, the Reformation showed a strong tendency not to reject medieval faith but to shift it to the Protestant values of seeking authority from the Bible rather than the Pope and his Vatican advisors. The rise of a middle class also led to an explosion of literacy made possible by Johannes Gutenberg (c.1398–1468) and the invention of affordable printed texts for all fields of knowledge past and present.

In the Enlightenment, values shifted again and deism became fashionable among scientists. They saw a clockwork universe with a dualism that kept the supernatural out of the day-to-day workings of the universe. Laws of nature, not everyday miracles, allowed the world of science to explore how that universe operated. Dualism was ascendant. Scientific societies formed. The Industrial Revolution shifted values again, and the world of industry leaned on science for new inventions, new fields of science, new products, new discoveries, and the wealth of products brought back by explorations and colonization. These are certainly revolutions of culture. They are less due to new theories reassembled from older theories than they are to the incremental effects of accumulating knowledge and expanding European presence around the world through conquest and trade. (We speak of the American Revolution and not the American paradigm shift.) The cultural parallel is not one of a geocentric model shifting to a solar model of the Sun, Earth, and planets. Monarchs had been deposed since antiquity. Democracies have precedents going back to Athenian Greece. Copernican theory (certainly a paradigm shift) became scientific astronomy when the telescope was introduced, allowing Galileo to show support for a model with detailed evidence that the geocentric system was less likely than one with material planets, including the Earth, orbiting a larger sun that rotated on its axis.

How Culture Influences Scientific Field Formation

In science, ideas change within the limits of such cultural climates. We cannot conceive of a cell theory before there were microscopes. We cannot conceive of a chromosome theory of heredity before there were advances in stain technology to make structures within the cells discernible. We cannot conceive of a molecular biology without a preexisting chemistry and biochemistry that made it possible. Each contemporary theory of living things was dependent on past knowledge of the biology of plants and animals. In all likelihood, the incremental changes will continue as each of these broad theories of life is challenged by new findings at the level of molecular biology and systems biology, the most recent of the fields of life. We also cannot predict the outcomes of emerging sciences, such as those studying the proteins emerging at different times in the life cycle or in different tissues of the organism. We do not presently know how the portion of living matter, the cytosol, in which known organelles are embedded, is organized and how it functions. Whether we call it cytosol or use its older terms, cytoplasm or protoplasm, these are imprecise words from the perspective of molecular biology. It would be vanity to believe we know from current knowledge how to make a cell in its entirety from a shelf of chemicals. At best, our technology allows us to replace the natural DNA of a prokaryote or eukaryote with a synthetic "off-the-shelf" constructed DNA that serves as a working chromosome in the altered bacterium or eukaryotic cell.[1] But that still does not reveal how the recipient "protoplasm" is organized and how it works in relation to the cell's organelles. In the past, this led scientists to invoke vitalism or holism. This is less likely today because the reductionist model has been powerful with the publication of tens of thousands of articles that have revealed exquisite detail on cells, chromosomes, genes, and cytoplasmic organelle structure and function.

Could one argue today, as Kuhn did some 40 years ago when I discussed paradigm shifts and biology with him, that the life sciences are largely descriptive and not theoretical? I counter that the five broad concepts—evolution by natural selection, the cell theory, the theory of the gene, the molecular basis of life, and the unfolding of the life cycle through generations—are very broad theories. Each has major consequences for how we look at our own lives and how we see the universe. What makes the life sciences unique is the diversity of its species and individuals within any species. Another feature is the ease with which experimentation can be performed in the life sciences. Finally, the life sciences invite the invention and use of abundant tools for analysis of the multiple levels of complexity of life from viruses to mammals. All of these attributes make new theories less necessary because, so far, these five major theories of biology have been well tested and have satisfactorily evolved in their complexity and sophistication.

I have not written a history of the life sciences. Nor have I given detailed attention to the many dozens of other figures who played a role in the formation of these fields and theories. This book is not a monograph. It is a sustained argument that the life sciences have few paradigm shifts at a major level of generating new theories by overthrow and reorganization of preexisting scientific theories. There are revolutions in the life sciences, but I believe my analysis shows that they are dependent on new tools and procedures that more frequently lead to new fields without overthrowing old ones. Microscopy, biochemical analysis, cell fractionation, breeding analysis, DNA sequencing, embryo manipulation (*Entwicklungsmechanik*), anatomical dissection, chromatography, centrifugation, and radioactive labeling are some of these tools necessary for new fields to emerge or old ones to change. They lead to fusions of cognate fields of knowledge. It is the desire to explore the unknown, to get nature to reveal its hidden organization and functions, that stimulates creative scientists. For those scientists, nature and science are not constructions but interpretations. It is not consensus but evidence that convinces them of their findings or the findings of their colleagues or even their competitors.

Distinguishing Paradigm Shifts from Incrementalism

Kuhn explored several philosophic issues associated with his model of revolutions occurring through paradigm shifts. He believed that consensus, not evidence, prevailed in the establishment of a paradigm shift. A major reason for this, he argued, was the incommensurable nature of the old and the new paradigms. Each dictated the relations of the components in different ways. Is the moon a satellite of the Earth in a solar system or is it a planet around the Earth in a geocentric system? The shift to a new paradigm comes from the simplicity of the new paradigm, which eliminates awkward anomalies like the retrograde motion of the other planets caused by the Earth's orbit around the sun. Kuhn's writings and examples are drawn from physics, in which Kuhn had the background for careful analysis. But his book was widely read in other sciences and especially the social sciences (psychology, political science, economics, and sociology). The social sciences are limited in the type of experiments they can do. They also deal with complex interacting agents that make a simple reductionism impossible. The life sciences have shifted from descriptive science (anatomy and physiology) to experimental science (resulting in fields like genetics and embryology). The life sciences also shifted to biochemistry and molecular biology largely through the tools making these fields possible (centrifugation, chromatography, isotope markers, X-ray diffraction). In the life sciences, tools generated theories, not the reverse. This makes it easier for a classical geneticist (cytogenetics and breeding analysis) to talk with a biochemical geneticist (e.g., George Beadle's biochemical pathways in fruit fly eye color inheritance) or a molecular biologist (e.g., the gene's specificity of information is associated with the aperiodic sequence of nitrogenous bases).

Controversy Does Not Necessarily Lead to Paradigm Shifts

Some biologists have doubted the existence of paradigm shifts. Adam S. Wilkins in 1996 raised this question in his *BioEssays* editorial. He argued that natural selection, the double helix model of DNA, and the introduction of Mendelism in 1900 did not fit that model.[2]

As the 19th century faded and the 20th century began, three scientists took issue with the prevailing model of heredity favored by Darwin's advocates. They had accepted the unlikelihood of Darwin's theory of pangenesis with its gemmules shed by cells and assimilated by reproductive cells. They had shifted from an implied Lamarckian acquired inheritance in that model to a model proposed by Weismann with a constancy of the germplasm in relation to the soma or body cells. The body modifies itself in response to the environment but the germplasm does not. For the biometricians, as they called themselves, heredity resided in the combinations of units of inheritance which participated in statistical ways yielding bell-shaped curves for traits.[3] Bateson countered this theory with numerous examples of meristic (duplication of parts) and homeotic (displacement of parts) mutations that produced discontinuous variation.[4] The biometricians argued these were of no evolutionary value and consisted of "monstrosities" and pathologies. Eventually Bateson's views were assimilated into the new synthesis, and individual gene mutations that effected developmental processes became part of evolutionary biology.

de Vries also took issue with the rigidity of the biometric school. He published his experimental findings that new species could arise in the genus *Oenothera*.[5] He believed that there was no fine line between fluctuating variations of the Darwinian sort, single gene variants likely to be encountered by breeders, and massive alterations of traits as he encountered in the new species of *Oenothera* that arose en masse. Only a portion of de Vries' theory survived because most of his findings were reinterpreted when classical genetics in fruit flies revealed some basic mechanisms relating genes, chromosomes, and evolution. By the 1930s, de Vries' findings had been reduced to polyploidy and recombination within sets of translocated chromosomes that rendered many recessive traits homozygous. Many of the permanent hybrids he found were associated with balanced lethals, an interpretation that fruit fly genetics offered.[6]

The third group to take on the biometric model was Morgan's fly lab. They showed that sex chromosomes existed and had a modified Mendelian pattern of inheritance for X-linked genes. They showed these genes could be mapped. They demonstrated the occurrence of nondisjunction to produce aneuploid strains of fruit flies. They demonstrated gene-to-character associations by experimental analysis isolating chief genes from their genetic and environmental modifiers. They supplemented Darwinian gradualism with classical genetics. They distinguished quantitative traits from discontinuous traits and eventually showed how biochemical processes, organ formation, body plan symmetry, and other aspects of biology could be assimilated by the chromosome theory of heredity.[7]

In all three cases, no paradigm shift was involved. Controversy is natural to all fields of science and each new finding leaves critics free to point out weaknesses and new ways to test the strength of interpretations of data. Each new method (breeding analysis, cytological study of chromosomes in aneuploids) generates more data and more convincing interpretations and excludes some unlikely models such as the universality of statistical curves for all character traits of evolutionary value.

Similar arguments could be made for the controversies associated with the stability of the gene. Castle took issue with it when a Mendelizing trait for hooded rats or spotted rabbits suggested that the genes were producing the pied patterns. Muller's work on beaded wings and truncate wings showed to the contrary that "residual inheritance" or modifier genes were the major factors involved in shifting the expression of the chief gene for these traits.[8] Castle eventually recognized this in his own mammalian experiments. What worked was not a paradigm shift in thinking but an effective experiment to test contending models. Contending models are numerous, if not universal in science. This is as true for breeding analysis (classical genetics and its numerous controversies) as it is for the chemical basis of heredity, which shifted from nuclein (a nucleoprotein) to nucleic acid in the 1890s, with sperm as largely vehicles bearing a nucleus of nucleic acid. By 1915, this view had shifted back to proteins as having a major role in being the hereditary material as enzymes began to assume a major role in biochemistry and as nucleic acids turned out to be relatively simple crystals with an inferred "tetranucleotide" repeating pattern in the studies of Phoebus Levene (1869–1940). It shifted back to nucleic acids with the work of Oswald Avery and his Rockefeller colleagues using DNA as the transforming substance that provided the genes extracted from killed bacterial cells to express themselves in a recipient host bacterial cell. Here it was the skilled purification of components and appropriate enzymes to digest them that gave Avery's experiment the edge in interpreting the chemical basis of heredity.[9] Alfred Hershey (1908–1997) and Martha Chase (1927–2003) soon confirmed and extended that analysis in their bacteriophage experiments, which used radioactive labeling to detect viral DNA and viral protein function.[10] What works in biology, more often than any other approach, is the use of new tools or procedures to show support or contradiction for an interpretation based on experimentation, not paradigm shifts.

The Components of New Field Formation and Scientific Progress

For the life sciences, I would argue that the primary components for forming new fields of science are (1) the discovery of relatedness among isolated components and (2) integrating apparent contradictions. The first of these I would compare with the parable of the five blind people and an elephant. Each investigator experiences a tail, leg, ear, trunk, or wall of flesh. As they converse and argue and each is allowed to move to the other's place, they soon discover that the five experiences are explained as components of a single animal, an elephant. It cannot be shuffled by a paradigm shift into a giraffe or a rhinoceros because the components of those animals would be experienced differently. The elephant would have a unique interpretation from any five persons willing to discuss their experiences and invite each other to the same place. Thus, chromosomes in the 1890s are leading to meiosis and a possible chromosome theory, but it is incomplete. In 1900, breeding analysis reveals Mendel's laws. Seemingly these components are unconnected. But Wilson, Sutton, and Boveri contributed to a chromosome theory of heredity in which chromosomes differ in their hereditary activities and recombining chromosomes gives different genetic outcomes. The chromosome theory of heredity thus combines two isolated approaches—cytology and breeding analysis—and it is then called genetics by 1906.[11] It will add other components in the next two decades, including population genetics and evolution. Eventually it will include biochemistry, developmental genetics, and molecular biology.

The second component of incrementalism is the integration of apparent contradictions Kuhn called anomalies. This process is what Kuhn designated as normal science. Using another parable, I would liken it to the familiar argument that one cannot compare apples and oranges. A good biological example would be the evolutionary implications of chromosomes and genes. Chromosomes through nondisjunction or mitotic failure can lead to speciation by polyploidy in many plants and to speciation by aneuploidy in many plants and some animals. Gross chromosomal rearrangements have a potential to serve as isolating mechanisms for evolutionary divergence, and many species reveal inversions or translocations that arose and shifted populations of a common species into two or more new species. Chromosomes also enable evolutionary change by repeat formation (tandem duplications) associated with unequal crossing-over. Many species harbor transposable elements inserted into the genome; kindred species may have very few such transposable elements in their genomes. Also, some rearrangements result in altered functions of genes called position effects, which do not actually change the sequence of the responding gene. This evolutionary list of chromosomal effects is different from the list of genic events associated with evolution, which include frameshift mutations, transitions, and transversions at the molecular level. They do not deal with the functional aspects of genes identified as amorphs (no normal function), hypomorphs (reduced normal function), neomorphs (novel functions expressed by an altered gene), and hypermorphs (excess expression of a trait). Nor do the chromosomal approaches explain homeotic mutations or meristic genes associated with body plan development. They do not account for gene nests or complex loci resulting in pseudoallelism or multiple allelism or genetic fine structure. They do not explain operons or regulatory genes. They do not account for introns, exons, and polymorphisms within and between species. They cannot account for genetic complementation in a pseudoallelic series.

One important difference between the two parables and the two examples I cited is that they describe activity associated with a moment in time. In the incrementalism model, the finding of a tusk might take place a century after the finding of a tail. In the life sciences, fields may emerge over centuries and the components involve new tools and new data pouring in.

The two approaches to gene and chromosome involvement in evolution are integrated through the genome itself. Both gross and subtle rearrangements play roles in evolution. So do gene mutations of various sorts and the relation of genes to one another through regulated activity in metabolism or regulated activity in embryonic development. The analysis of a genome at the nucleotide level can be used to identify past rearrangements of all types and the signatures or sequences of nucleotides can reveal gene function.[12] The contradictions shift from cytology versus gene studies through breeding analysis to DNA studies that would have been difficult to interpret without the knowledge of Mendelism, the chromosome theory, and cytogenetics that preceded molecular analysis of DNA.

Are There Paradigm Shifts in the Life Sciences?

If the five major theories that constitute the life sciences are not paradigm shifts in Kuhn's original sense or replacements of competitive paradigms that became outmoded, are there paradigm shifts at some other level of study? Medicine has shifted in speculative ways since antiquity. Some, like Galen, saw illness as an imbalance of vital humors. Some believed the body was made toxic by agents that could be purged by vomiting or eliminated through enemas. Some believed those toxins could be reduced in quantity by opening a vein and removing some of the blood. Some believed there were miasmas or contagious odors that caused illness. Some physicians who were also religious believed lack of faith caused illness and restoring that faith brought about cures. Some believed people varied in their constitutions, which made them vulnerable to diseases. What these all lacked was evidence.

The first scientific efforts to identify the causes of contagious diseases were performed by Ignaz Semmelweiss (1818–1865) in Vienna and by Oliver Wendell Holmes, Sr. (1809–1894) in Boston, who recognized the importance of a thorough cleaning of the hands and change of clothes before approaching a patient's body for examination. A generation later, the cause of infectious diseases was identified not as "morbid particles" but as germs or bacteria by Louis Pasteur. Both Semmelweiss and Holmes had evidence to support their findings, and Pasteur and later Robert Koch had solid microscopic evidence for the presence or absence of specific microbes associated with specific infectious diseases. Just as Copernicus was preceded by Aristarchus for a solar system model, these 19th century biologists and physicians were preceded by Girolamo Fracastoro in Renaissance Padua who proposed a theory of invisible spores as the agents causing infectious diseases. Pasteur had the insight to use the spoilage of wine and extend it to putrefaction of foods and then to the cause of specific contagious diseases. In each case, he relied on his microscopic observations and carefully controlled experiments.

If we measure scientific revolutions by impact, the germ theory of infectious diseases ranks high. If impact is the measure of a paradigm shift, then the germ theory is properly such a shift in medical training and understanding. But impact tells us little about the methodology of science or how the concept or field evolved.

Both Kuhn's model of paradigm shifts and the model of incrementalism presented here share a belief that scientific progress or development fits a model of evolution. For Kuhn, that evolution is not directed to a goal. Each paradigm shift arose to resolve anomalies that could not be dismissed in the old paradigm. Incrementalism also involves the emergence of new fields (e.g., the cell theory lead to the chromosome theory which lead to the theory of the gene). In the incrementalism model, I suggest, it was not anomalies that caused the new fields to emerge. It was the introduction of new tools and new fields that were complementary in their interests. They could easily use a shared vocabulary built out of the old and new findings. If I were to use an example, I would cite Cuvier's and Agassiz's model of successive acts of extinction and a new flora and fauna replacing the old as one interpretation of life on earth. Implied in that catastrophe model was a supernatural Creator. Quite different was Darwin's view of the history of fossils and contemporary life, which evolved in complexity and diversity over long periods of time. Darwin saw chance variations, chance changes in environments, and isolating cofactors as agents leading to evolution. Darwin's model won out because it provided testable ways to distinguish the two causal theories. The fossil record supports it, comparative anatomy supports it, geographical isolation supports it, and molecular phylogeny supports it. This is not consensus. This is evidence-based argumentation characteristic of experimental and descriptive sciences that are data driven more than they are theory driven. If Kuhn's evolutionary model of the progress of science is an evolutionary one, it is certainly not a Darwinian one.

One could argue that catastrophe theory has returned with meteor impact or massive volcanism as models for past extinctions. But contemporary catastrophe models do not involve total replacement of old phyla with new phyla, and there are survivors in each of the new geological eras that were not wiped out by the planetary event. Evolution is not an either/or mutual exclusion model with Creationism as its alternative. Evolution is a composite of many fields and findings. Darwin recognized this in his 20-year wait before publishing his book and would have waited even longer had he not been forced to publish when Wallace sent him a near-identical theory.

Krishna Dronamraju (b. 1937) in 1989 tried to rescue the model of paradigm shifts by creating a collection of differentiating terms—major paradigms, minor paradigms, subparadigms, co-paradigms, incipient paradigms, delayed paradigms, extinct paradigms, hybrid paradigms, and many others.[13] The difficulty I saw with this effort was that it made everything one does in science some sort of paradigm. When a term becomes so inclusive that it covers all aspects of scientific interpretation it loses its original meaning. What is important in Dronamraju's analysis is his recognition that science involves many processes of thinking, comparing, experimenting, classifying, and assigning validity. Can I think of a major field of the life sciences that is comparable to what Copernicus and Galileo did in shifting the geocentric universe into the solar system? No. Evolution by natural selection is probably of comparable significance to the Copernican revolution. But as Darwin formulated it, the theory of natural selection was heavily scientific in its evidence (geographic isolation, evidence from domesticated breeds, fossils, isolating mechanisms, comparative anatomy, comparative embryology). What it displaced (Creationism, special creation, fixity of species from species arising by some unexplained act or acts of creation) was largely grounded in a religious tradition found in the book of Genesis or in miraculous acts of gods in polytheistic religions. The science in creationism, then and now, is miniscule.

Candidate Accomplishments May Be Revolutionary but They May Not Qualify as Paradigm Shifts

There are several great contributions to the life sciences that some identify as paradigm shifts. These are (1) Weismann's theory of the germplasm, (2) Charles Darwin's evolution by natural selection, (3) Barbara McClintock's "jumping gene" model of the dynamic genome, (4) the Watson–Crick double helix model of DNA, and (5) Crick's proposal of the "genetic dogma" or DNA makes RNA makes protein. Both a paradigm model and an incrementalism model would claim these as scientific revolutions. Each of these, as these chapters have shown, has a history.

Weismann and his contemporaries knew that there were two ways to look upon heredity. Some believed like Lamarck that acquired characteristics were inherited and this had evolutionary implications. Others recognized the passage of mutant or varietal characteristics over generations without change and believed many traits were innate. Weismann did not live in a world where all scientists were Lamarckian in their views of heredity. They were divided. Weismann amassed evidence from circumcision and head and foot binding in humans to animal studies, including his own study of mutilated mice tails to bolster the view that there was a separation of the germinal material from the somatic changes taking place in the life cycle. He also relied on his dissertation research that showed in *Diptera* there was a formation of a "polar cap" into which future germ cells migrated in the developing embryonic egg and were set aside for reproduction. I believe this disqualifies his revolutionary finding as sharing the same conceptual changes (the shift in paradigm shift) as Copernicus.

I do not consider Darwin's proposal of evolution by natural selection in Kuhn's 1962 sense. There were precedents for evolution provided by Jean-Baptiste Lamarck, Darwin's grandfather Erasmus Darwin (1731–1802), and others going back to ancient Greek speculations (Democritus and Lucretius). The language of Darwin's theory was commensurate with Creationists like George Cuvier and Louis Agassiz (1807–1873). Before 1858, biologists objected to speculative evolution (e.g., Robert Chambers' [1802–1871] *Vestiges of the Natural History of Creation*, 1844). Darwin's evidence was abundant and made use of geology, comparative anatomy, comparative embryology, domestic selection, geographical diversity, and the fossil record. All of these fields were known to Creationists and shared a common professional vocabulary.

Darwin was well aware of Lamarck's contributions to the ideas of an evolution of life. He was also aware of the controversy over Robert Chamber's *Vestiges of Creation* and the uproar this book caused (Chambers published it anonymously). He believed it was unproved speculation, like Lamarck's evolution. Darwin amassed evidence. He used a Baconian approach to let the connections emerge from a huge accumulation of information initiated by his voyage on the *Beagle*. Even that did not generate his model until he came back and after reading Malthus came up with the idea of a competition to survive and a natural environment as the selective agent among a diverse population. Darwin's chief obstacle to publishing was the widespread religious view that life and the universe had to be created by a Creator and could not arise from natural processes. His opposition was not a scientific outlook but a supernatural outlook that wanted

God to play a role in either the creation of species or the guidance of that process of evolution. In the Copernican model of a paradigm shift, God is not involved in the mathematics or physics of the objects of the universe. Their roles switch. A planet (the Ptolemaic moon) becomes a satellite. The Sun shifts from being a planet around the Earth to being a star. The Earth becomes demoted to a planet orbiting the Sun. There is no role switching in Darwin's theory of evolution by natural selection. He offers a causal explanation for his observations instead of a supernatural one.

McClintock's finding of jumping genes was worked out in maize. She was curious about a pattern of streaks and dots in the kernels of corn she studied. She made the interpretation that some genetic element was capable of being detached and inserting elsewhere in the maize genome. Such new associations turn the activity of neighboring genes on or off. This was a major discovery. It also turned out to be a complex process. Some species have many transposable factors that are inserted in their genomes. Other species, like the puffer fish, have very few such inserted elements in their genomes. McClintock had hoped she would work out an orchestration of these movable genes and construct how a life cycle emerges from this interplay. Even if her hope turns out to be correct, it still does not explain why the puffer fish has so few transposable genes that Sydney Brenner offered it as a candidate for the earliest effort to work out a vertebrate genome in the first years of the Human Genome Project. The difficulty I see in the historical and philosophical interpretation of how science works are the varied meanings of the term "paradigm shift." It has become a synonym for the term "scientific revolution." It is used as a synonym for a great finding, experiment, or theory and thus is measured by its impact rather than by its methodology.

The Watson–Crick model of DNA was a great finding and the competition among Pauling's laboratory, the Wilkins–Franklin laboratory, and the Watson–Crick Cambridge laboratory reflects how keenly aware all three were of the importance of working out a structure of DNA. DNA had for the previous ten years been the center of attention for the role as the chemical basis of heredity. Caspersson and Schultz showed the correlation of mutagenicity and the ultraviolet wavelength that damaged DNA. The pneumococcal transformation studies of Avery's Rockefeller group showed DNA was the transforming substance. Hershey and Chase used isotope labeling to show the roles of DNA and protein in virus growth. DNA was clearly the source for generating the viral life cycle. There was no groundswell of experimentation to show that proteins played such a hereditary role. There were lots of theoretical reasons for thinking of the gene as being a chemical that had a capacity for storing specificity (or information) and even producing enzymes, as the work of Beadle and his colleagues showed. Muller's early insights into the gene as a crystal-like molecule that could copy its variations and serve as the basis of life were part of a widely held view that Watson and Crick shared. Schrödinger's *What Is Life?* revealed the biological properties that the double-helix model provided. Measured by impact, the double-helix model was as significant a finding as Darwin's proposal of evolution by natural selection. It was revolutionary as basic science in biology shifted to molecular biology as the most exciting field to enter. But that is a measurement of impact. It is not a contribution to a theory of paradigm shifts. If the model of paradigm shifts has a specific model (and Kuhn did provide that model), then that specific model of the philosophy of science should apply to the working out of the double helix, but that's not how it happened. X-ray diffraction was used. Model building was used. Chargaff's biochemical findings were used. Muller's 1936 characterization of the gene as a type of crystal reproducing its variations was used. The procedures are very different from the renaming and relocating and reassignment of roles that gave birth to Kuhn's interpretation of the Copernican revolution as a paradigm shift.

Crick's central dogma, "DNA makes RNA makes protein," I would call a working model of how genes work. The double-helix model, as its investigators pointed out in a follow-up note to *Nature*, predicted the major activities of genes and hence the DNA that now composed them. Crick predicted there would be an intermediate molecule (RNA and later more specifically messenger RNA) that would carry the information from the DNA (genes) to the cytoplasm in which metabolism occurred. He also predicted adaptor RNA (later transfer RNA) to carry the amino acids to the messenger RNA. Zalokar showed evidence of support for Crick's central dogma using radioisotopes to locate RNA production, first in the nucleus and later that same activity of the labeled RNA ended up in the cytoplasm of what later was identified as the ribosomes. The components of the process for decoding the messenger RNA sequence of nucleotides and for the translation product (a sequence of amino acids) was incrementally placed in the emerging model of protein synthesis in the endoplasmic reticulum of the cell cytoplasm by the ribosomes within it processing the messenger RNA through a complicated process that was duplicated in test tubes.

All five of these candidates for paradigm shifts in the life sciences fail, I believe, to abide by the standard used by Kuhn for a paradigm shift. None of them arose from anomalies that could not be explained by contemporary models. None of them involved a shifting of roles, terminology, or broad mental picture guiding the scientists working in that field. There was no incommensurable barrier between the ideas and vocabulary of those who made the revolutionary contribution and those who read it with either resistance or pleasurable acceptance. What they share in common is that they are revolutions because they led to new fields of the life sciences and they had enormous impact. I also believe that the incrementalism is a more faithful interpretation because of the numerous pieces that came together to shape the mental picture that emerged in each case.

What Is the Significance of an Incrementalism Approach?

I do not doubt the validity of Kuhn's paradigm shift interpretation of the Copernican revolution. The fields of chemistry, geology, and psychology may find similar examples. Nor do I argue that the incrementalism I find in the life sciences must apply to other sciences, although I would not be surprised that experimentation and new tools and techniques play significant roles in generating new fields in these broader branches of science. I am not convinced that the life sciences are purely descriptive and lack a theoretical biology. Overwhelmingly, most experimental biologists reject vitalism, a mode of thinking about life that has

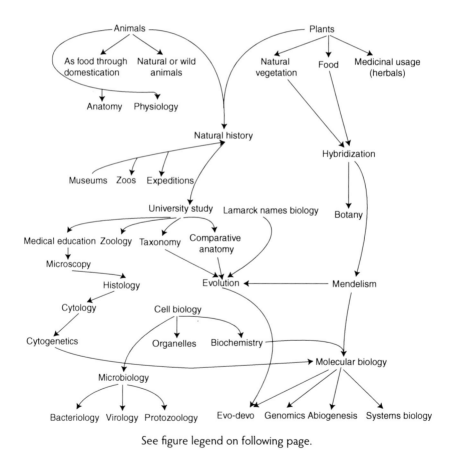

See figure legend on following page.

been largely replaced by reductionism. I am also aware that much of biological science remains unexplained. Biologists cannot provide a detailed account of what constitutes the cytosol of the cytoplasm—that part of the cytoplasm not designated as an organelle—of a cell whether eukaryotic or prokaryotic. At present, that responsibility has been shifted to the growing fields of "omics," which explore genomes, proteomes, transcriptomes, and other dynamic or interacting epigenetic processes in the living cell. At the DNA level, the major problems seem more connected and fall into a reductionist's framework. J. Craig Venter's (b. 1946) use of "off-the-shelf" synthesis of a bacterial genome that works in a cell whose DNA has been destroyed is fairly convincing to reductionists. But there is nothing comparable for the molecular basis of components, organization, and function of the cytosol.

This does not bother me because I recognize it has taken some three billion years of evolution to produce life as we know it and it is overconfidence to believe that in our generation all remaining major aspects of life sciences can be solved. As a historian of science, I appreciate what my predecessors did working with limited tools. I am confident that the study of the thousands of gene differences in comparative genomics will reveal new relations and processes in the coming

---

Animals and plants were the two groups of life recognized since antiquity. Animals were seen as sources of food (especially if domesticated) and as wild or natural. A similar division for food and natural vegetation existed for plants, but a third area of interest arose for medicinal properties of plants. The study of animals quickly led to two divisions. Dissection was the primary tool for anatomy. Studying function was the province of physiology. Collectively, the study of life belonged to the field of natural history. Museums, zoos, and expeditions dominated natural history during the Renaissance and Enlightenment. A separate interest arose in hybridization by practical breeders and scholars studying classification. Studies were performed in museums, universities, herbaria, gardens, and zoos. Medical education and natural history were overlapping, especially for botany, which was a significant portion of medical education. The introduction of the microscope spawned histology and then cytology and dominated the field of pathology in medicine. Comparative anatomy and taxonomy strongly influenced the field of evolution first by Lamarck and then by Darwin. The fields of microbial biology arose through the germ theory and differentiated into bacteriology, protozoology, virology, and even more splintered fields of phycology, mycology, and cryptogamic botany. A similar splintering occurred in zoology with vertebrate and invertebrate biology splintering into primatology, herpetology, entomology, and other specialties. By the 19th century with the Ph.D. as a research degree, an abundant specialization took place with tools for each specialty. The electron microscope, the centrifuge, chromatographic separation of molecules, radioactive labeling, and X-ray diffraction all playing parts in bringing molecular biology out of biochemistry and physiology. Today's life sciences stress evo-devo, comparative genomics, abiogenesis, and systems biology as areas in which much is not known and in which research is likely to produce novel ideas and findings if not new fields of the life sciences.

generations to generate interpretations of cell dynamics and composition as well as spinning off new fields and perhaps new theories of the life sciences.

References and Notes

1. Gibson G, Glass JI, Lartigue C, Noskov VN, Chuang RY, Algire MA, Benders GA, Montague MG, Ma L, Moodie MM, et al. 2010. Creation of a bacterial cell controlled by a chemically synthesized genome. *Science* **329**: 52–56.

2. Wilkins AS. 1996. Are there "Kuhnian" revolutions in biology? *BioEssays* **18**: 695–696.

3. Schwartz J. 2008. *In pursuit of the gene: From Darwin to DNA*. Harvard University Press, Cambridge, MA.

4. Bateson W. 1894. *Materials for the study of variation*. Cambridge University Press, New York. Also see Cock AG, Forsdyke DR. 2008. *Treasure your exceptions: The science and life of William Bateson*. Springer, New York.

5. De Vries H. 1902. *The mutation theory*. Open Court, Chicago.

6. Carlson EA. 1966. *The gene: A critical history*. Saunders, Philadelphia.

7. Carlson EA. 1966. Op. cit.

8. Castle WE. 1915. Mr. Muller on the constancy of Mendelian factors. *Amer Nat* **49**: 37–42. Muller HJ. 1918. Genetic variability, twin hybrids, and constant hybrids, in a case of balanced lethal factors. *Genetics* **3**: 422–499.

9. Olby R. 1974. *Path to the double helix: The discovery of DNA*. University of Washington Press, Seattle.

10. Hershey AD, Chase M. 1952. Independent functions of viral protein and nucleic acid in growth of bacteriophage. *J Gen Physiol* **36**: 39–56.

11. Wilson EB. 1902. Mendel's principles of heredity and the maturation of the germ cells. *Science* **16**: 991–993; Sutton WS. 1903. The chromosomes in heredity *Bio Bull* **4**: 231–251; Boveri T. 1902. On multiple mitoses as a means for the analysis of the cell nucleus. *Verh Phys Med Ges Würzberg* **35**: 67–90 (translated into English in 1968. *The chromosome theory of inheritance: Classic papers in development and heredity* [ed. Voeller B], pp. 87–94. Appleton-Century-Crofts, New York.)

12. Gregory TR. 2005. *The evolution of the genome*. Elsevier, New York.

13. Dronamraju K. 1989. *The foundations of human genetics*. Thomas, Springfield, IL.

There Is More to Scientific Revolutions Than Paradigm Shifts and Incrementalism

Science versus pseudoscience, debatable science, and fraud. Scientific personalities, contrarians, ad hominem arguments in science. Politics and science.

In this argument concerning paradigm shifts and incrementalism as models for scientific revolutions, new field formation, and progress in the life sciences, I have assumed the integrity of both the scientists involved in research and the safeguards that keep science from making errors. But such an ideal state of science is unrealistic. In this appendix, I present concerns about the nature of science that reflect fraud, wishful thinking, denial of science, conflicts of interest (especially in applied science), and other factors that make scientific objectivity and public trust in science difficult.[1] None of these is directly involved in the debates on paradigm shifts and incrementalism. They are more likely to delay progress in science or deflect science into the issues of fraud, the invoking of the supernatural, and the politicizing of science.

What Distinguishes Science from Pseudoscience?

Good science is open to criticism and can be tested for its claims and predictions. Karl Popper called this the falsification test of science. Every experiment opens the possibility that an accepted theory might be false in its predictions. Science requires publishing in peer-reviewed journals or books. Papers in which evidence is lacking or interpretations are not supported by the data or experiments cited are sent back to the authors for revision. Error is very common in science because people make mistakes, the materials used in different laboratories may differ in purity, the environmental factors differ in different laboratories, the organisms used in the life science experiments may vary in genotype, and the tools may differ in efficiency. Good results, however, stimulate repetition and extensions of a theory or a finding that is published. This leads to much improved descriptions and outcomes of a developing science. This type of precision is chiefly associated with science since the late Renaissance. Before that collection of activities (using controls, testable hypotheses, introducing better tools to obtain data, and later on statistical analysis), called the scientific method, existed, science varied in quality. In the absence of such tools and habits, interpretations were often guesses.

Science Sometimes Emerges from Pseudoscience

Astrology long assumed that wandering planets, their location during the seasons and among the fixed stars, and their constellations were opportunities for predicting events. Sky gods, angels, or heavenly devices (crystalline spheres) were invoked to account for these movements and for hidden messages that astrologers could interpret. Lucky astrologers were rewarded. Many others cited errors but did not lose faith in what they were taught about astrology. When I visited the Jagiellonian University in Kraków in 2001, there was a large mural in the school of medicine foyer depicting Copernicus (a physician as well as an astronomer) and the zodiac signs and their relation to health. Physicians drew a horoscope for each of their patients in the 15th and 16th centuries. Astrology failed the controlled tests of its predictions once statistical equations could be applied to the comparison of known data and incidences and the predicted incidences of different astrologers.

Also more than two millennia old is the field of physiognomy. Facial features were associated with personality traits. The nose, eyes, lips, ears, and chin were scanned for their association with both virtuous and antisocial traits. It nearly cost Darwin a trip on the *Beagle* (Captain Robert FitzRoy did not like the shape of Darwin's nose). It was made into a pseudoscience by Cesare Lombroso (1835–1909), who believed criminals had fixed features that he observed while doing autopsies on executed prisoners. He had a strong influence on the development of degeneracy theory. Lombroso and his followers believed the criminal body type and facial features reflected an atavism to an ancestral type with brutish behavior. Among the traits he singled out was a desire for tattoos, a lust for blood and mutilation of victims, and a low intelligence.[2]

Similar conjectures were made to justify augury (the study of liver anatomy to predict the future), palmistry, tarot card readings, mind reading, table rapping (by ghosts or spirits), and phrenology (the belief that bumps and depressions in the skull reflected personality and talents). More sophisticated pseudoscience involved racial typing, particularly by Count Joseph Arthur de Gobineau and Houston Stewart Chamberlain (1855–1927). The eugenics movement applied its findings to social failure, mental incompetence, psychosis, and personality. These attributes were then applied to social class, national origin, and race. They led, in the United States, to the eugenics movement's compulsory sterilization laws and restrictive immigration acts and to religious and racial bigotry.

What characterizes these social pseudosciences is their appeal to the vanity of one group (usually middle class or wealthy white Protestants) and their use of science to justify their prejudiced assumptions.[3]

Some Fields of Science Are Controversial

Some fields are controversial in their status, with scientists inclined to consider them as pseudosciences and nonscientists more open to their validity. This is particularly true in the health fields. Chiropractic medicine is recognized in

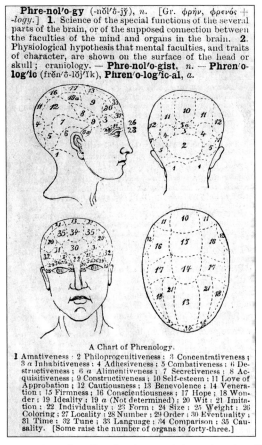

Phre-nol′o-gy (-nŏl′ŏ-jў), *n.* [Gr. φρήν, φρενός + *-logy*.] **1.** Science of the special functions of the several parts of the brain, or of the supposed connection between the faculties of the mind and organs in the brain. **2.** Physiological hypothesis that mental faculties, and traits of character, are shown on the surface of the head or skull; craniology. — **Phre-nol′o-gist**, *n.* — **Phren′o-log′ic** (frĕn′ŏ-lŏj′ĭk), **Phren′o-log′ic-al**, *a.*

A Chart of Phrenology.

1 Amativeness ; 2 Philoprogenitiveness ; 3 Concentrativeness ; 3 *a* Inhabitiveness ; 4 Adhesiveness ; 5 Combativeness ; 6 Destructiveness ; 6 *a* Alimentiveness ; 7 Secretiveness ; 8 Acquisitiveness ; 9 Constructiveness ; 10 Self-esteem ; 11 Love of Approbation ; 12 Cautiousness ; 13 Benevolence ; 14 Veneration ; 15 Firmness ; 16 Conscientiousness ; 17 Hope ; 18 Wonder ; 19 Ideality ; 19 *a* (Not determined) ; 20 Wit ; 21 Imitation ; 22 Individuality ; 23 Form ; 24 Size ; 25 Weight ; 26 Coloring ; 27 Locality ; 28 Number ; 29 Order ; 30 Eventuality ; 31 Time ; 32 Tune ; 33 Language ; 34 Comparison ; 35 Causality. [Some raise the number of organs to forty-three.]

This dictionary entry on phrenology lists the characteristics assigned to different regions of the skull with the underlying assumption that bumps, size, and deformities of these regions reflect human behaviors. Franz Joseph Gall was an early neurologist and some of his findings have survived; however, phrenology was based on the self-deception that he could read character by examining the surface of the skull's cranium. Today, phrenology has become the prototype of pseudoscience.

most states of the United States as a valid alternative medicine. Lots of patients feel benefitted for back pain after spinal manipulation. Few scientists accept the thesis of chiropractic medicine that nervous fluids are unblocked by manipulation or that most diseases stem from blocked passage of fluid in the vertebral column. Similarly, few scientists would accept that the virtual elimination of detectable medication is what makes homeopathic prescriptions effective when taken by patients. In these medical fields, patient testimony is offered as evidence. Few controlled studies have been performed to test the success ratios of mainstream medicine and these alternative medical programs.

Freudian psychiatry constructed talk therapy sessions with underlying assumptions about sexual conflicts as the basis for neurotic and psychotic behavior. Freudian approaches fell into disfavor in the last half of the 20th century and psychiatrists turned to medication as a way to control behaviors. Many psychiatrists are still impressed by the way our minds reflect Freudian insights. Freudian slips in conversation and erotic thoughts in non-erotic settings are virtually universal experiences. Although oedipal models may lack scientific rigor, the phenomena Sigmund Freud explored, such as sublimation of discontents, fears of castration, or shifting one's own dysfunctional personality traits onto others, still require attention from psychiatrists for both diagnosis and treatment.

Acupuncture has been shown to be effective in Chinese medical practice for some surgeries. Herbal medicines are still used in many countries and some of

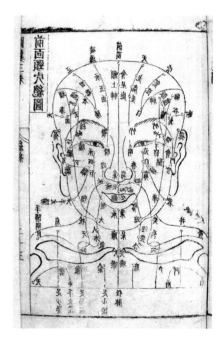

Acupuncture is still used in China and many other countries in the world as a form of alternative medicine. Instead of anesthesiology, patients may prefer the apparent numbing effect of needles placed at certain parts of the body to prevent pain of surgical or other medical procedures. Whereas many anesthesiologists consider acupuncture a form of quackery, others are more open to limited usages and, unlike phrenology, which is universally rejected as pseudoscience, acupuncture remains one of the debated fields of applied science. The sites and regions that acupuncture needles control claimed to have come from arrow wounds of warriors in China's past. That aspect of acupuncture is most likely to be considered as pseudoscience.

those medicines have been extracted, synthesized, and applied to orthodox treatments. A major difference in orthodox medicine is the long and expensive process that drug companies have to employ to assure that a new product is both safe to use (does no harm) and effective (actually prevents or cures or treats the disease).

For basic science, there are not as many questionable fields as in medicine. One such field is based on the Gaia hypothesis by James Lovelock. The Earth is seen as an organism that adjusts to changes associated with volcanism, disturbances in weather patterns (storms, droughts, ice ages), or the arrival of epidemics and pandemics. The homeostasis of the body is applied to the Earth as a whole. Critics of the Gaia hypothesis argue that ecological findings, evolutionary mechanisms, and geological theories are sufficient to explain changes and maintenance of climate, species diversity, species extinctions, and species migrations in different habitats (islands vs. continents). Critics also believe that the Gaia hypothesis appeals to those seeking a supernatural explanation of life and the universe.[4]

Personality, Ideology, Culture, and Self-Interest Influence How Science Works

Most scientists are skeptics. H.J. Muller said of his mentor T.H. Morgan that "he doubted the doubt until he doubted it out." Morgan preferred experiments to generating hypotheses. Muller believed both were essential for science, and he was not afraid to seek connections from the data of his experiments. Some scientists use polemics or ad hominem arguments when discussing the work associated with a competitor.[5] Journal editors have mostly avoided such comments in publications of peer-reviewed articles. Most scientists would agree that a hypothesis should stand on its own testable merits. Some scientists experience conflicts of interest. They may be funded by pharmaceutical companies, the tobacco industry, nuclear energy suppliers, forestry companies, seed suppliers to farmers, or the military industries. How objective is their applied or basic research when that funding source has a special interest in the findings of the scientists they support? Many journals require a statement at the end of the published article that no such conflict of interest exists or they do list the organization or organizations that supported the research reported.

Besides skepticism, scientists deal with other personality traits. Some like to challenge well-established beliefs or claims of settled issues in science. They are called contrarians by their critics. Richard Goldschmidt and William Castle in genetics were contrarians. They rejected modifier genes as the basis of fluctuations in expression of a Mendelian genetic trait. They rejected crossing-over among chromosomes as a means to constructing linear genetic maps. Goldschmidt saw the gene as similar to a finger pressed on a violin string. The notes were of the whole string, not separate genes like piano keys in his thinking. Both

Castle and Goldschmidt made good contributions to genetics, but they also forced geneticists to spend time designing additional experiments to support the findings contrarians dismissed as wrong. Contrarians and zealous skeptics often invoke Galileo as their model of science. Unfortunately, so do medical quacks and those who commit fraud. Frauds are usually caught because it is difficult to anticipate how nature works if you do not do the experiments to find out how things work. Second guessing nature is overwhelmingly likely to lead to false rather than correct interpretations. Reputation in science is critical for acceptance of one's work or obtaining grant support.

Much more difficult to recognize or control in a scientist's career is the role of culture. Religion plays a major role in some scientists' lives. Before there was a germ theory, plagues were considered a visitation or punishment sent by God to chastise a population for its sins. Children born with birth defects were seen as omens of state importance. Some believed birth defects were a consequence of forbidden sexual unions. Francis Galton challenged the concept of public prayer (said at the dinner table while saying grace and praying for the health of the royal family). He compared the mean age of male life expectancy, the causes of death, and the frequency of illnesses in the royal family with families of well-to-do non-royals such as lawyers, doctors, scientists, artists, and literary figures. He found no major difference in age at death (royals achieving 64 and the eminent non-royals varying from 64 to 67, the highest being physicians). He concluded that public prayer either was ineffectual or was rejected by God as insincere.[6] Most religions include supernatural beliefs, especially the existence of one or more gods and assorted spirits, souls, or divine manifestations in the universe. Scientists are

Rudolf Virchow was a German physician and filled with contradictions. He believed the cause of excess illnesses among the laboring class was due to low wages. He believed cancer had its origin from a single cell. He is the founder of cellular pathology in medical schools. He proved by study of all of the German states that there was no such thing as an Aryan race. He popularized and named the cell doctrine, although he later acknowledged the idea for it came from Robert Remak. One of his most successful students was Ernst Haeckel, who became an ardent proponent of Darwinism in Germany for which Virchow called him a fool. Virchow did not believe evolution had taken place on Earth. He considered his religious outlook as that of an agnostic. He founded public health programs, including free health examinations for children, but did not believe in the germ theory of epidemic diseases. He fits the classification of a contrarian personality whose good, fortunately, outweighed his bad judgment.

sometimes dualists and compartmentalize the supernatural aspects of their religious tradition and keep it separate from their scientific habits.

Nationalism may play a role in science as it did in the USSR during the 1930s to the 1960s, when state ideology decided what was acceptable to Marxist philosophy and what was rejected as bourgeois, idealistic, antisocial, or a threat to Soviet society. The Lysenko controversy was about genetics and the environment.[7] A similar nationalism in the USSR was invoked to reject some of Einstein's theories as philosophically at odds with materialism. For a short time, Soviet embryologists revived the free formation of cells and rejected the cell doctrine. Hitler and his Nazi movement made sure that racial propaganda (especially Aryan superiority and anti-Semitism) was presented as legitimate science.

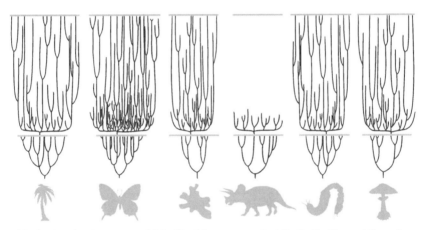

Inside the Creation Museum, exhibits like this one argue that the Earth, life, and the universe arose by special creation by God more than 6000 years ago. During creation week, all life-forms were created by God according to their "kinds." Representatives of the land animal kinds were on Noah's Ark and those that remained outside the ark were wiped out 4350 years ago by the flood described in the Old Testament. The surviving kinds differentiated into the varieties (the present species) over this relatively short period compared to the millions of years associated with evolutionary phylogenetic trees. In the museum, the "kinds" are typically defined as orders or families.

Author's note: The legend above provided by the Creation Museum keeps the definition of "kinds" vague (taxonomic Orders or Families). Such changes would be at least as dramatic as changes from ape-like primate ancestors to our present *Homo sapiens*. The evidence provided by evolutionists described in Chapter 12 uses comparative anatomy, the fossil records, geographic distribution, cytogenetic correlation, and molecular sequence comparisons. Greek and Roman scholars 2500 years ago tried to explain sex determination or the causes of contagious diseases with guesswork. So, too, religious documents of that era could not provide evidence for their interpretations of the origins of life in the absence of sophisticated tools, abundant data, or controlled experiments.

Big Science and Little Science

Before the 20th century, most scientific papers had a single author. That began to change with the publications in genetics of Bateson and his colleagues, whose reports to the evolution committee were coauthored. A similar multiauthored policy emerged with the publications of Morgan and his students in the fly lab. Morgan was appreciated for this democratic policy that recognized the contributions of students as coauthors. As the complexity of science grew in the 20th century, the costs of science increased. Physics led the way, especially during World War II, with research on radar and the development of the atomic bomb. After the war ended, Vannevar Bush successfully argued for a National Science Foundation in the United States to support basic science education and experiments. This was extended to the National Institutes of Health. Big science was rare before that and had found its expression in organized expeditions, such as those that were formative to the careers of Huxley and Darwin in their extensive studies of life in little-visited regions of the planet. In physics, studies of atomic particles required expensive construction of cyclotrons, linear accelerators, colliders, and other devices, with dozens of coauthors on each paper reflecting the numerous tasks in the design, execution, and interpretation of data. A similar effort and expense came to the life sciences with the Human Genome Project, a $2.7 billion program involving many universities and research institutions that required about 15 years to complete.[8]

How Science Works

The various pseudosciences, disputed sciences, and varieties of personal and cultural factors discussed in this appendix often lead to debates about how science works. They suggest that the conflict between paradigm shifts and incrementalism is only part of a broader picture of how science works. Objectivity, integrity, and an opportunity to do supported research are essential components for scientists to flourish. For nonscientists sorting out what is valid science from wishful thinking, contrarian personalities, conflicts of interest, and supernatural belief is difficult to assess. Although many scientists teach courses for nonscience majors, not all students accept a scientific worldview of life and the universe. This creates crises for funding research, defending scientific research, and evaluating scientific claims for legislation.

References and Notes

1. Carlson E. 2006. *Times of triumph, times of doubt: Science and the battle for public trust.* Cold Spring Harbor Laboratory Press, Cold Spring Harbor, NY.

2. Lombroso C. 1899. *Crime: Its causes and remedies*. Little, Brown, Boston.

3. Carlson E. 2001. *The unfit: A history of a bad idea*. Cold Spring Harbor Laboratory Press, Cold Spring Harbor, NY.

4. Lovelock J. 2000. *Gaia: A new look at life on Earth*. Oxford University Press, Oxford; also see Gould S. 1997. Kropotkin was not a crackpot. *Nat Hist* **106**: 12–21.

5. Carlson E. 2017. Scientific feuds, polemics, and ad hominem arguments in basic and special interest genetics. *Mut Res* **771**: 128–133.

6. Galton F. 1872. Statistical inquiries into the efficacy of prayer. *Fortnightly Rev* **12**: 125–135.

7. Soyfer V. 1994. *Lysenko and the tragedy of Soviet science*. Rutgers University Press, New Brunswick, NJ.

8. Hilgartner R. 2017. *Reordering life: Knowledge and control in the genomics revolution*. MIT Press, Cambridge, MA.

An Incrementalist Timeline
of the Cell Theory

This timeline illustrates the major features of the cell and gives a sense of how the components of the cell and their functions were worked out since the cell was named in 1665. We start with an empty box and end up with a membrane-bound unit with sophisticated internal organelles distributed to the outer cytoplasm and a set of chromosomes within the bounds of the nuclear envelope. Neither the terms "normal science" nor "paradigm shift" apply to these stages of the timeline narrative of incrementalism. Key concepts like mitosis or meiosis, cellular metabolism, or organelle functions are absent in Hooke's time. They did not replace earlier ideas of cell function. The only function Hooke suggested was buoyancy.

1665   *Robert Hooke* describes cork tissue as composed of cells. He uses a microscope he made and sees empty boxes and associates them with buoyancy and lightness of weight.

1670–1690   *Antonie van Leeuwenhoek* uses a higher-powered lens to describe animalcules he isolates from many environments.

1675–1679   *Marcello Malpighi* uses microscopy to reveal an anatomy that includes capillaries to account for the shift from arterial to venous blood.

1797   *Francois Xavier Bichat* describes organs as composed of tissues. He classifies 21 different tissues using dissection with needles rather than microscopic examination.

1830   *Joseph Jackson Lister* eliminates spherical and chromatic aberration of lenses by preparing achromatic lenses composed of two different layers of chemically distinct glass (with or without added lead). The idea of achromatic lenses was first proposed by *John Dollond* and *Chester Hall* in 1758.

1833   *Robert Brown* describes a spherical region in some plant cells that he calls a nucleus.

1835–1869   Cells are identified as fluid-filled and the names sarcode (*Felix Dujardin*, 1835) or protoplasm (*Hugo von Mohl*, 1846; *Max Schultze*, 1854; and *T.H. Huxley*, 1869) are used to describe them.

1838   *Matthew Schleiden* describes all plant tissues as composed of cells.

1840   *Theodor Schwann* describes all animal tissues as composed of cells.

1840   Both *Schleiden* and *Schwann* believe cells form by free formation like crystals in a liquid.

1855   *Robert Remak* and *Rudolf Virchow* propose a cell doctrine that all cells arise from preexisting cells. Virchow introduces a field of microscopic pathology to medical students. He argues cancers arise from a single-cell origin.

1857   Stain technology is introduced by *Joseph von Gerlach* using carmine as a dye to bring out cell structure.

1875   *Walther Flemming* describes mitosis as a process involving chromosome distribution.

1876   *Otto Bütschli* describes fertilization as a union of one sperm and one egg.

1876–1883   Meiosis is worked out by *Oscar Hertwig* and *Edouard van Beneden* and the concepts of haploid, diploid, and constancy of the chromosome number are introduced.

1890   Mitochondria are described by *Richard Altmann* and *Carl Benda*.

1931   The electron microscope is introduced by *Ernst Ruska*.

1945–1955   The electron microscope describes organelles of cells: cell membrane, nuclear envelope, mitochondria, lysosomes, and Golgi apparatus. *George Palade*, *Albert Claude*, and *Christian de Duve* are major contributors.

1950–2000   The biochemistry and molecular biology of cell organelles are worked out using X-ray diffraction, microbial genetics, radioisotope autography, cellular fractionation, chromatography, centrifugation of components, microsurgical manipulation of cell parts, and immunological tools.

Figure Credits

Cover: Flemming W, 1882. *Zellsubstanz Kern und Zelltheilung.* Verlag von F.C.W. Vogel, Leipzig. Reproduced courtesy of Lilly Library, Indiana University, Bloomington, Indiana.

Chapter 1: *Page 8*, Bill Pierce/Time Life Pictures/Getty Images.

Chapter 2: *Page 12*, Micrographia; page 13, reprinted from Leewenhoeck A. 1684. *Phil Trans* **14:** 568–574; *page 18*, Wellcome Collection; *page 22*, reprinted from Flemming W. 1882. *Zellsubstanz Kern und Zelltheilung.* Verlag von F.C.W. Vogel, Leipzig; *page 23*, reprinted from http://speculativeevolution.wikia.com/wiki/File:Primitive_Eukaryote_Cell.jpg.

Chapter 3: *Page 31*, reprinted from Sturtevant AH. 1913. *J Exp Zool* **14:** 43–59; *page 34*, reprinted from Sturtevant AH. 1925. *Genetics* **10:** 117–147; *page 36*, courtesy of Caltech, *Engineering & Science* magazine.

Chapter 4: *Page 46*, reprinted from Bateson W. 1894. *Materials for the study of variation treated with especial regard to discontinuity in the origin of species*, p. 287. Macmillan, London.

Chapter 5: *Page 64*, reprinted from https://commons.wikimedia.org/wiki/Category:Francesco_Redi#/media/File:Redi_Francesco_1626-1697.png; *page 66*, redrawn from https://en.wikipedia.org/wiki/Swan_neck_flask#/media/File:Louis_Pasteur_Experiment.svg; *page 68*, redrawn from https://commons.wikimedia.org/wiki/File:0332_Cell_Cycle_With_Cyclins_and_Checkpoints.jpg.

Chapter 6: *Page 79*, courtesy Ava Helen and Linus Pauling Papers, Oregon State University Libraries; *page 81*, reprinted by permission from Springer Nature, Watson JD, Crick FHC. 1953. *Nature* **171:** 964–967; *page 82*, reprinted from Benzer S. 1959. *Proc Natl Acad Sci* **45:** 1607–1620; *page 83*, reprinted by permission from Springer Nature, Crick F. 1970. *Nature* **227:** 561–563.

Chapter 7: *Page 94*, reprinted from Stevens NM. September 1905. Carnegie Institution of Washington, Publication No. 36, Plate VI; *page 96*, reprinted from https://ghr.nlm.nih.gov/gene/SRY; *page 97*, reprinted with permission from Dr. Steven M. Carr, © 2008 by Terra Nova Genomics, Inc.

Chapter 8: *Page 107*, reprinted with permission from the Genetics Society of America, Crow JF. 1998. *Genetics* **148:** 923–938; *page 109*, reprinted from Muller HJ. 1918. *Genetics* **3:** 424S; *page 114*, reprinted from Davenport GC, Davenport CB. November 1910. *The American Naturalist* V.44, No. 527, 641–672.

Chapter 9: *Page 118*, reprinted from https://ro.wikipedia.org/wiki/Girolamo_Fracastoro#/media/File:Fracastoro.jpg; *page 120*, original photomicrographs by Robert Koch, Public Domain; *page 121*, redrawn from https://socratic.org/questions/how-are-the-lytic-and-lysogenic-cycles-different; *page 125*, reprinted from https://phil.cdc.gov/phil/details_linked.asp?pid=20854, Cynthia Goldsmith.

Chapter 10: *Page 132*, reprinted from https://en.wikipedia.org/wiki/Daphnia#/media/File:Daphnia_pulex.png, photo by Paul Hebert, 2005. *PLoS Biol* V. 3/6; *page 133*, reprinted from

Hartsoeker N. 1694. *Essai de Dioptrique*; *page 135*, reprinted from Roux W. 1890. *Über die Entwicklungsmechanik der Organismen.*

Chapter 11: *Page 148*, reprinted by permission from Springer Nature, Zalokar M. 1959. *Nature* **183:** 1330.

Chapter 12: *Page 156*, reprinted from https://en.wikipedia.org/wiki/Carl_Linnaeus#/media/ File:Linnaeus1758-title-page.jpg; *page 158*, redrawn from https://commons.wikimedia.org/ wiki/File:Voyage_of_the_Beagle-key.svg; *page 161*, (*left*) reproduced by kind permission of the Syndics of Cambridge University Library, classmark: Tree_of_Life.tif (DAR121:36), (*bottom*) reprinted from https://commons.wikimedia.org/wiki/File:Phylogenetic_tree_scientific_names. svg; *page 162*, reprinted from Bridges CB. 1936. *Science* **83:** 210–211; *page 163*, reprinted from Hinchliff CE, Smith SA, et al. 2015. *Proc Natl Acad Sci* **112:** 12764–12769, with permission from Stephen Smith.

Appendix 1: *Page 187*, reprinted from https://en.wikipedia.org/wiki/Phrenology#/media/ File:1895-Dictionary-Phrenolog.png; *page 188*, reprinted from https://commons.wikimedia .org/wiki/File:Acupuncture_chart_of_front_of_head,_17th_C._Chinese_woodcut_Wellcome_ L0034707.jpg, Welcome Library reference: External Reference Wang Shumin 17-2, External Reference Chou 41/1624 Ma 17-2 and External Reference Vivienne Lo; *page 190*, reprinted from https://en.wikipedia.org/wiki/Rudolf_Virchow#/media/File:Rudolf_Virchow_NLM4.jpg; *page 191*, image courtesy of CreationMuseum.org.

Flowcharts: Flowcharts designed by Christina Carlson.

Index

Page numbers followed by an f denote a figure.